CC 75 S95 1962
3 0050 00030 9434

D1555165

# Archeological Chemistry

## A SYMPOSIUM

# Archeological Chemistry

## A SYMPOSIUM

Edited by Martin Levey

Philadelphia
University of Pennsylvania Press

Published in Great Britain, India, and Pakistan
by the Oxford University Press
London, Bombay, and Karachi

Library of Congress Catalogue Card Number: 64-24497

7465
Printed in the United States of America

# Foreword

In recent years archeological chemistry has made great progress through the work carried out in the laboratories of many countries. The present collection brings much of this widely scattered work together and thus should be very welcome to those who are active in this field and also to many others.

If, nonetheless, the question were raised whether these "others" really can afford to spend any time with this subject when so much is required and remains to be done in the pursuit of our present problems, the answer would be quite positive and diversified. First, interest in this subject has deep historical roots. Several outstanding chemists of the eighteenth and nineteenth centuries studied ancient objects. Berthollet, Chaptal, and Vauquelin analyzed ancient pigments. Humphrey Davy, who pointed out that azure had shown its stability over 1700 years, derived practical suggestions from his work on ancient Roman coloring materials, for example, the use of "artificial hydrat of alumina" as a binder for pigments. Davy leads us to part of the answer, for which we can rely on Louis Alphonse Salvétat, director of the national manufacture of porcelain at Sèvres (1850). He saw the "incontestable value" for "our arts" that can be gained from such research, and its more general interest in helping us to "reconstitute the ancient chemical arts." From this we can see the greater perspectives this work offers to general history when it is concerned with commerce and communications, with the origins and migrations of techniques. And all this touches on the riddle of creativity insofar as it is connected with the problems of beginnings and influences.

There is a feedback between answers and questions. If my answer were accepted, it would lead to the further question of whether archeological chemistry can do all these things, especially, whether we can elucidate what went on so long ago in the "art" we now call chemistry. While trying to make out the chemistry of the ancients, we are developing a modern chemistry of ancient specimens. We are measuring radioactivity, we use the electron microscope and spectroscopy in its various forms, but the results we obtain with these modern tools must be carefully distinguished from what the ancients did and how they may have arrived at their achievements without anything like the knowledge acquired in the course of the intervening centuries. How much of the ancient product has remained "frozen" in its original condition? How much change has occurred through self-diffusion, recrystallization, and external influences?

Here we encounter a new example of complementarity. Where our study fails to show the original nature of the ancient material, it can succeed in revealing new relationships in long-term chemical changes. Either we find the ancient process or we discover the history of its product.

This interplay of questions and answers can serve to indicate the many directions in which archeological chemistry as a field of specialization is connected with other inquiries and why it should interest not only the specialist, but many others besides.

Washington, D. C. EDUARD FARBER

# Editor's Preface

The present collection of studies in archeological chemistry brings together the most important group of articles ever devoted to the difficult experimental problems in this field. Not only do the chapters have their own individual importance but they have a collective value as the work of some of the foremost scientists in this rapidly growing field.

Organized by the undersigned, the Third Symposium on Archeological Chemistry was held at Atlantic City, New Jersey, on September 12 and 13, 1962. The official sponsor was the Division of History of Chemistry, of the American Chemical Society. It was the major purpose of the chairman to bring together the latest knowledge and techniques of archeological chemistry so that more and better investigation in this area might receive recognition and the proper encouragement.

Close to 90 percent of the major types of the investigative methods of archeological chemistry are described in this volume. Abrahams-Edelstein employ infrared spectra in the analysis of extracted dyes. In Besborodov-Zadneprovsky and in Mellichamp-Levey, emission spectroscopy is the primary experimental approach. Long series of analyses are given by Caley and Geilmann. The primary approach by Emoto is by a new method using X-ray fluorescent spectroscopy. Jedrzejewska uses, in addition to other methods, petrographic examination. Mellichamp-Levey also employ X-ray diffraction. Naumann uses metallographic, hardness, and analytical data. Examination by electronic sound and micrography are preferred by Panseri-Leoni. In discussing prehistoric sherds, Paramasivan considers color, thickness of fabric, hardness, specific gravity, porosity, thermal properties, properties determined optically, and analyses. The C-14 dating technique is given by Ralph. Sayre-Smith and Yamasaki concentrate on chemical analyses, and Weill works with such X-ray techniques as fluorescence and diffraction. One technique, not discussed but which may prove itself of importance, is that of neutron bombardment (described by E. V. Sayre in "Studies in Ancient Ceramic Objects by Means of Neutron Bombardment and Emission Spectroscopy," *Application of Science in the Examination of Works of Art* (1959), 154–155.

During the course of the discussions on each of the presented papers, much attention was devoted to the specific methods of laboratory work, significance of the results, and the terminology of the reports. Because the representation of talks is of a truly international geographic character, the differences in actual approach, although of great value in some ways, clearly indicate that archeological chemists must develop an identifiable standard of reporting in their analytical work.

Research in archeological chemistry is still largely conducted by chemists and physicists working alone in their own particular specialties. Usually, this is done without a full recognition and appreciation of the possibilities of study using other available techniques. This is partly due to the relatively few workers in the field and also to the diluted communication of results which are spread out over many different types of publications throughout the world. A greater spirit of cooperation among archeological chemists in consultation and exchange of artifacts as well as symposia and specialized publications will help to overcome many of the present difficulties.

Beyond matters of experimental technology, the use of the data by the participants tends, in some areas, to reveal a maturing of the science of archeological chemistry. With a further improvement of statistical methods, of technique and its utilization, it is hoped and expected that this growth in the archeological domain will spill over its know-how into more modern problems in ceramics, metals, and other materials.

The papers are arranged alphabetically by author in this volume.

The chairman wishes to thank the American Academy of Arts and Sciences for aid which eventually contributed to the success of this undertaking. Appreciation of support by the U.S.P.H.S. RG 7391 is acknowledged by the undersigned. Thanks are also due to Drs. Earle R. Caley, Eduard Farber, and Henry M. Leicester for acting as presiding officers of the three sessions.

Yale University, New Haven, Conn.
Martin Levey, Chairman
Committee on Archeological Chemistry
American Chemical Society
December 11, 1963

# Contents

# Archeological Chemistry

## A SYMPOSIUM

# A New Method for the Analysis of Ancient Dyed Textiles

By David H. Abrahams and Sidney M. Edelstein,
Dexter Chemical Corp., New York

These authors were charged with the identification of the dyes present in eleven woven fabrics and in one sample of dyed unspun wool from the Bar Kochba finds in the Judean desert. The eleven fabrics were all wool and have been identified by Professor Yadin as clothing or shrouds used by the Bar Kochba rebels who retreated into a cave in the Judean desert in 135 A.D. The loose wool was found in a leather bag together with other household objects in the same cave. The fabrics and unspun wool were in an excellent state of preservation, and the colors were relatively bright and fresh looking after mild cleaning.

The methods used heretofore in the identification of dyes in ancient dyed fabrics have been based for the most part on qualitative color changes when the dyed samples were treated with various reagents.[1] Some analysts have used methods involving separation of the dyes by solvent extractions followed by attempts at identification of the dyes by examination of visible spectra of the dyes in solvent solutions.[2] We have felt that none of the techniques used heretofore have been exact enough for our purpose since they are subject to many errors and depend mainly on the skill and experience of the analyst. We have, therefore, developed a more positive method of identification which involves selective extractions of the dyes by solvents followed by purification of the dyes and then positive identification of each dye by comparison of its infrared spectra with the spectra of known dyes.

The Bar Kochba textiles were all wool and the colors were fast to mild washing. This, together with the age of the textiles, indicated a limited number of possible dyestuffs that could have been used in producing the colors.

From ancient times up until the discovery of America, fast reds were produced on wool with an extract of madder (alizarin) or with an extract of kermes or similar red insects together with alum as the mordant. Duller shades of reds, browns and some violets were obtained from these same dyestuffs by the use of iron salts in combination with the alum. Fast blues were obtained by the use of indigo from woad (*Isatis tinctoria*) or from the true indigo plant (*Indigofera tinctoria*). Purple shades were obtained by the use of true Tyrian purple (6,6′ dibromindigo) or by dyeing the wool blue with indigo and shading with madder or kermes in combination with an alum mordant. Fast yellows were available from weld, safflower, saffron and from many other plants in conjunction with an alum mordant. Fast greens were obtained by first dyeing the wool blue with indigo and then with one of the yellows together with an alum mordant. Variations in the green shades could also be made by changing the relative proportion of aluminum to iron in the mordants. Blacks were usually prepared by the use of an iron salt mordant together with an extract of tannin from oak galls or with other tannins. Occasionally, blacks were prepared by a combination dyeing of blue, red and yellow.[3]

## Analytical Procedure for Dyes in Fast-dyed Wool Textiles

*Scour.* Before analytical work is carried out, it is essential that the textile be cleaned. Each sample is individually washed in a warm solution of neutral soap followed by thorough rinsing in warm water. Agitation is held to a minimum because of the possible fragile nature of the textile. Each sample is then extracted in a Soxhlet apparatus with carbon tetrachloride, ethanol and distilled water and then dried.

The sample is dissected and examined under a low power microscope to determine if the wool in the particular sample has been dyed different colors and then blended to give the final color or whether the wool has been dyed one color only, either in the form of stock, yarn or in the piece. Depending on the results of this examination, the yarns or fibers may be separated into individual color groups for subsequent analyses.

*Determination of Metal Salts (mordants).* Each sample is destroyed by refluxing with concentrated nitric acid. The acid solution is then diluted and any residue removed by filtration. Qualitative analyses for the metals present in the solution or residue can be made by standard methods. When metals are present, their amounts can be estimated or exact quantitative determinations can be made by standard methods.

## Extraction of Dyes

*Carminic Acid, Related Color Acids and Safflower Yellow.* The dyed sample is immersed for 15 minutes in a minimum amount of warm dilute (3N) hydrochloric acid. This breaks the color lake and dissolves the metal mordant. Following this, the sample is washed with three portions of warm distilled water. The washings are then combined with the acid.

The combined solutions are evaporated to dryness on a steam bath, the residue dissolved in alcohol and evaporated again. This is repeated several times to remove last traces of free hydrochloric acid. Finally, the residue is dissolved in water and precipitated by the addition of a solution of lead acetate acidified with acetic acid. This precipitate contains carminic and related color acids. The filtrate will now contain safflower yellow if this is present in the textile. The precipitate is separated by filtration and thoroughly washed with water. The filtrate and washings are combined and held for determination of safflower yellow. The precipitate is then suspended in water and decomposed with a small amount of dilute sulfuric acid. The lead sulfate is filtered off, and the filtrate is evaporated to dryness. The residue is then dissolved in alcohol and filtered to remove any remaining metallic sulfates. The alcoholic solution is then evaporated and the residue is re-dissolved in water and precipitated again with lead acetate to form a crystalline lead salt of the carminic or related acids.

The filtrate retained for the safflower yellow determination is made alkaline with ammonia and if safflower yellow is present, it will be precipitated as the lead salt together with metallic hydroxides. The precipitate is removed by filtration and thoroughly washed with water. It is then suspended in water and decomposed with a small amount of dilute sulfuric acid. Lead sulfate is filtered off, and the filtrate is evaporated to dryness. The residue is then dissolved in alcohol and filtered to remove any remaining metallic sulfates. The alcoholic solution is again evaporated and the residue is re-dissolved in water to which a drop of acetic acid has been added. An excess of neutral lead acetate solution is added and the solution is made alkaline with ammonia to precipitate the lead salt of safflower yellow. The precipitate is washed and dried.

*Alizarin.* Following the extraction with the hydrochloric acid solution, the textile sample is dried. It is then extracted with carbon tetrachloride to remove alizarin (the

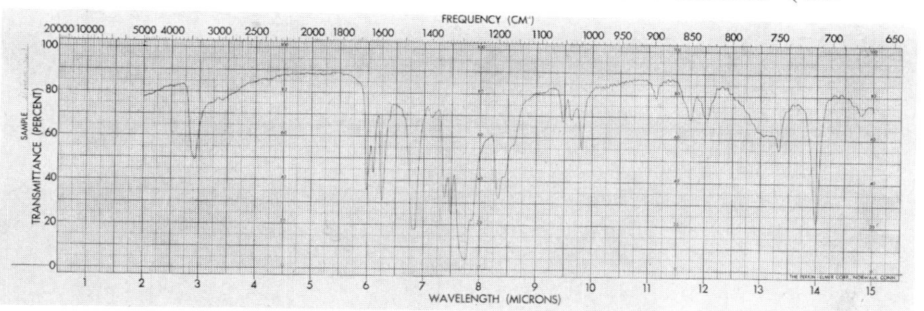


**Fig. 1. Synthetic Alizarin.**

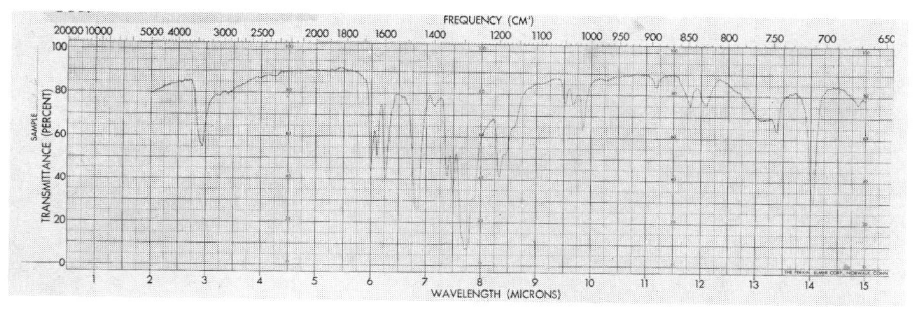


**Fig. 2. Alizarin Extract.**

coloring principle of madder). The carbon tetrachloride extract is evaporated to dryness and the extracted dye is purified by sublimation or by dissolving in 0.1 N. sodium hydroxide, then precipitating with 0.1 N sulfuric acid, filtering and drying.

*Yellow Dyes other than Safflower Yellow.* The textile from the previous two extractions is then extracted with warm ethanol to remove the yellow dyes such as those furnished by saffron, weld and other yellows. The yellow dye is then precipitated by the addition of alkaline lead acetate in ethanol. The precipitated lead salt is washed

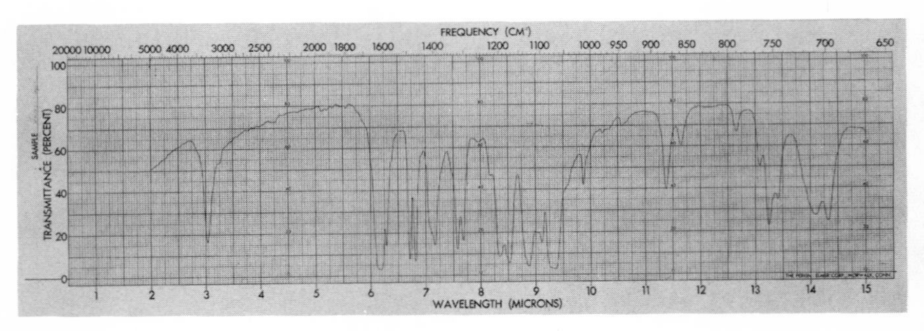


**Fig. 3. Synthetic Indigo.**

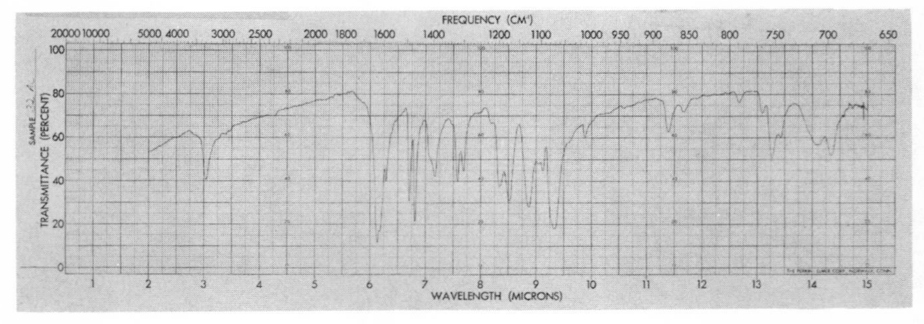


**Fig. 4. Indigo Extract I.**

thoroughly with ethanol and then with water. The precipitate is suspended in water and the lead salt is decomposed by the addition of hydrogen sulfide. After boiling to expel the excess hydrogen sulfide, the solution is filtered to remove the lead sulfide. The lead sulfide precipitate is washed with alcohol and this alcohol washing is combined with the water filtrate. The dye is again precipitated from the filtrate with alkaline lead acetate, washed with water and dried.

*Indigo.* The textile is next extracted with boiling glacial acetic acid to remove indigo. The acetic acid extract is evaporated to dryness on the steam bath. The indigo remaining in the evaporating dish is washed with water and with alcohol several times and then dried.

*Brominated Indigo.* The textile is finally extracted with a warm solution of alkaline sodium hydrosulfite to remove brominated indigo. The reduced dye solution is then oxidized with an excess of hydrogen peroxide and the precipitated dye is filtered off, washed with water and dried.

Note: While this method has been developed for fast dyes on wool, it is always possible that small amounts of brazil, orchil, barwood and many other dye materials might have been used for shading purposes on fast dyed textiles. Most of the dyes from these materials would be removed in the scouring procedure but small amounts might remain and would usually be extracted along with the yellow dyes. Tannins which were often used with iron salts for black shades would be separated with carminic acid.

*Infrared Examination.* Each purified dye or purified lead salt of the dye from the above extraction procedure is mixed with potassium bromide (spectrographic grade) and formed into a pellet or disc under high pressure for the determination of the infrared spectra of the dye. Each dye pellet is examined in a double beam infrared spectrophotometer between the wave lengths of 2 to 14 microns. The curves of these determinations are plotted by the instrument under the usual standard conditions. Infrared spectra are also prepared from known dyestuffs which had been subjected to the treatment used in the above extraction and purification methods for comparison with the unknown dyes extracted from the textile.

## Experimental Results

Positive identification is made by comparing the infrared spectral curves (spectrograms) of the purified extracts from the textiles with those of known dyestuffs. Figure 1 shows the infrared spectrogram for synthetic alizarin. Figure 2 shows the infrared spectrogram of alizarin extracted from a brown colored Bar Kochba fabric. It is obvious that the two curves are identical. Figure 3 shows the infrared spectrogram of synthetic indigo. Figure 4 shows the infrared spectrogram of the indigo extracted from an ancient green fabric and Figure 5, that for an ancient black fabric. Again, the three curves are identical. Figure 6 is the infrared spectrogram for 6, 6′ dibromindigo and is unquestionably different from the three indigo spectrograms. Similar results are noted in comparing the spectrogram for carminic acid extracted from cochineal (standard), Figure 7, with the spectrogram of the carminic acid extracted from an ancient purple wool, Figure 8. Figure 9 shows the infrared spectrogram of the yellow dye extracted from an ancient greenish-yellow wool fabric, and Figure 10 that of saffron. Figure 11 shows weld, Figure 12, Persian berries and Figure 13, safflower yellow. Only the spectrogram of

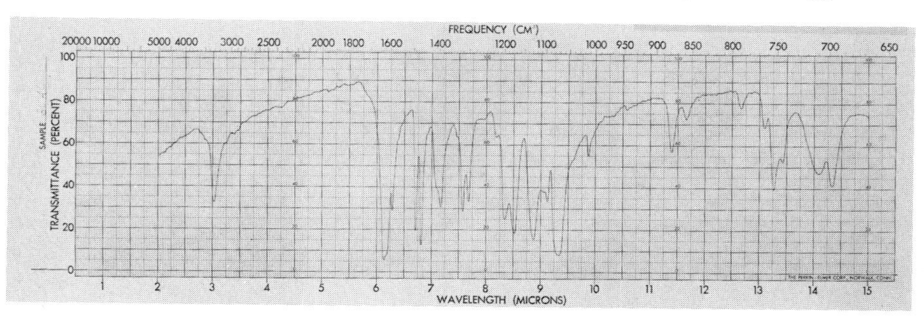


**Fig. 5. Indigo Extract II.**

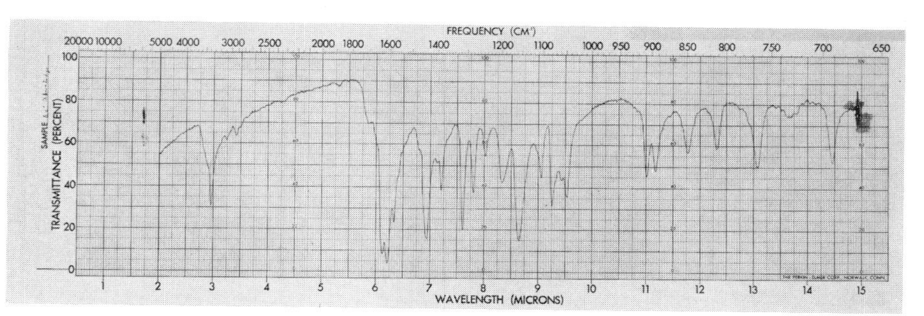


**Fig. 6. 6,6′ Dibromindigo.**

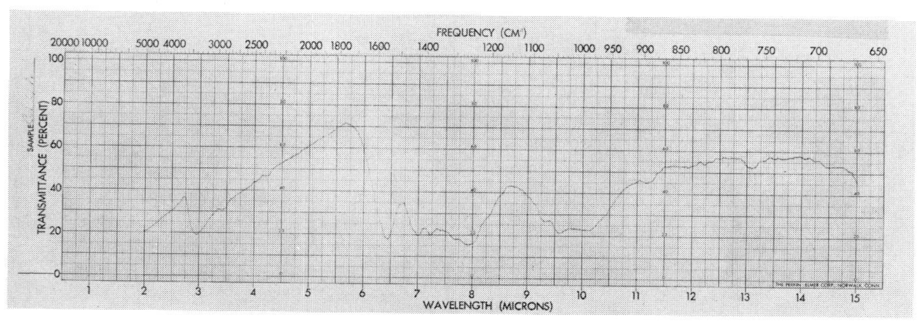


**Fig. 7. Carminic Acid Standard.**

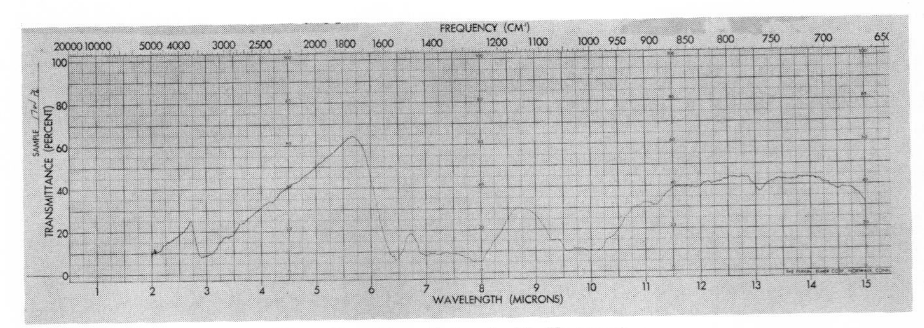


**Fig. 8. Carminic Acid Extract.**

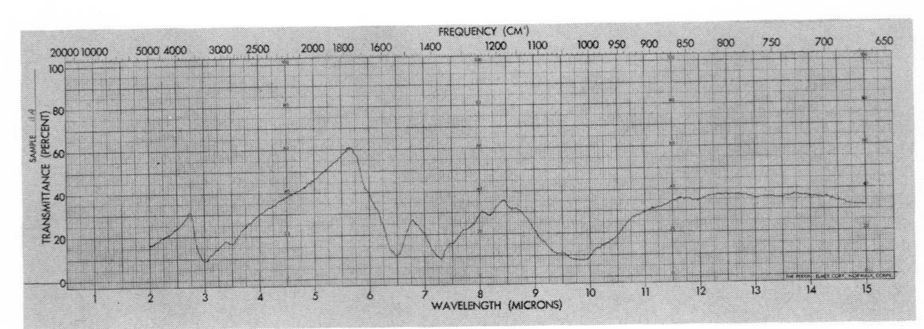


**Fig. 9. Saffron Extract.**

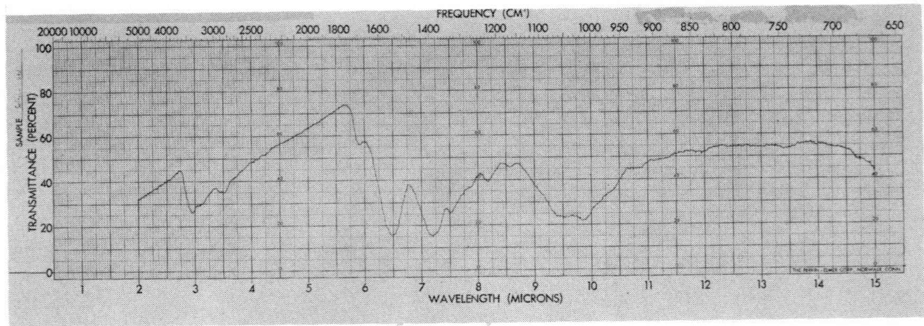


**Fig. 10. Saffron Standard.**

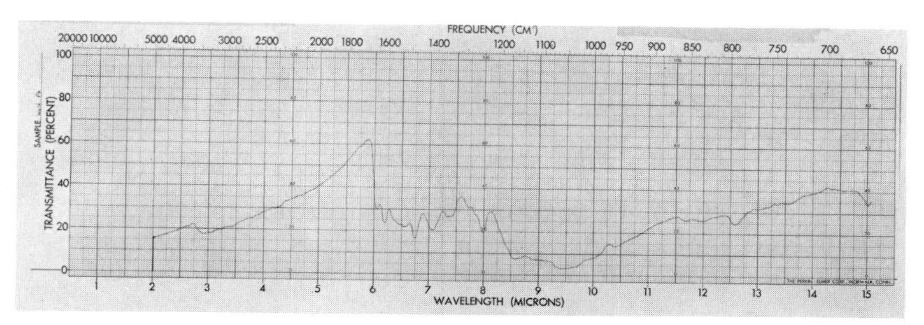


**Fig. 11. Weld Standard.**

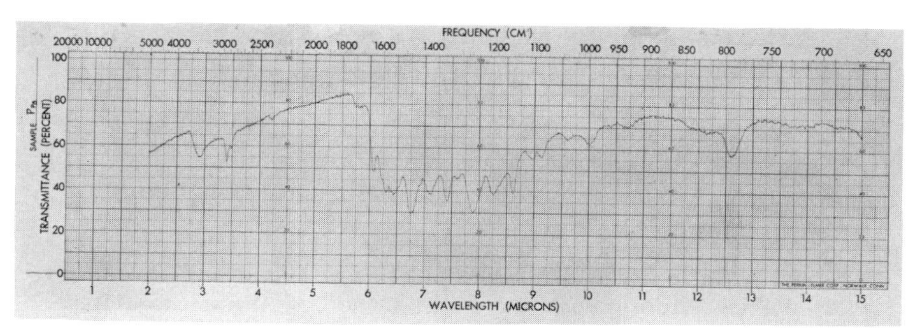


**Fig. 12. Persian Berries Standard.**

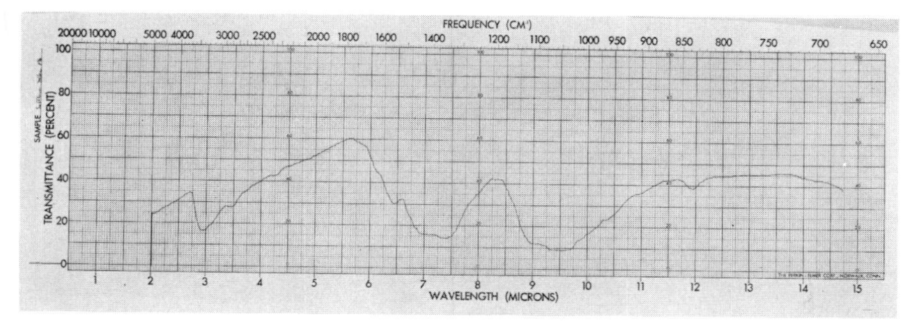


**Fig. 13. Safflower Yellow Standard.**

saffron (Figure 10) matches that of the yellow dye from the sample (Figure 9), and, therefore, the yellow dye in the fabric can only be saffron.

## CONCLUSIONS

The results of this work indicate that the method used is extremely satisfactory for the analyses of ancient fast-dyed wool textiles. The extraction method gives very sharp separations of the dye which normally would be encountered. In the case of yellows, however, there is always the possibility of difficulties if more than one yellow dyestuff had been used. Considerable skill would then be required in the examination of the infrared spectrogram of a mixture of different yellow dyes. It is also obvious that the method could be extended to the removal and analyses of dyes in ancient dyed cottons, linens and silks.

## ACKNOWLEDGEMENT

The authors wish to express their gratitude to Mrs. G. L. Shetky of the American Herb Society, to Dr. William J. Young of the Museum of Fine Arts of Boston and to Dr. Charles J. Weidmann of the Ciba Company for their help in obtaining the samples of natural and synthetic dyestuffs used in this study.

## NOTES

1. R. Pfister, "Teinture et Alchimie dans L'Orient Hellenistique," *Seminarium Kondakovianum* (Prague, 1935), VII, pp. 1–59. E. Knecht, C. Rawson and R. Lowenthal, *A Manual of Dyeing* (London, 1893), II, pp. 838–873.

2. Gustavo A. Fester and Siegfried Lexow, "Los Colorantes de la

Raiz de Relbunion Tetragonum de Cordoba," *La Revista de la Facultad de Quimica Industrial Y Agricola* (Santa Fe, Argentina, 1942–43), XI, XII, pp. 1–31.

3. R. J. Forbes, *Studies in Ancient Technology* (Leiden, 1956), IV, pp. 98–147. William F. Leggett, *Ancient and Medieval Dyes* (New York, 1944), pp. 1–91. C. L. and A. B. Berthollet, *Elements of the Art of Dyeing* (London, 1824), I, pp. 4–30. *Instruction Genérale pour la Teinture des Laines* (Paris, 1671), pp. 1–175. Edward Bancroft, *Experimental Researches Concerning the Philosophy of Permanent Colours* (London, 1813), I, II.

# Ancient and Medieval Glass of Middle Asia

By M. A. Besborodov and J. A. Zadneprovsky,
Leningrad

## Introduction

The earliest papers dealing with the history of glass-making in Middle Asia appeared at the end of the nineteenth century as a result of the well-known expedition of N. I. Veselovsky in 1884–1885 (70).* Excavations made by him on Toy-Tiube (Tashkent oasis) in Akhsikath and Uzgand (Ferghana) and on the Afrasiyab site in Samarkand uncovered important findings of glassware that stimulated wide interest in the development of glass-manufacturing in Middle Asia. Veselovsky published a paper based on his discoveries: "A Note on Glass Manufacturing in Middle Asia," (1894). This was an area which, according to a Chinese chronicle, demonstrated a high level of glass-making in the fifth century A.D.

Extensive development of archeological investigations, especially in the post-war years, has produced an accumu-

* Numbers in parentheses refer to the list of References at the end of the article

lation of important materials on the history of glass in Middle Asia. During this period, a series of articles dealing with the description of glass products were published.

A short review of the Middle Asian glass studies is given in the book, *Glassmaking in Ancient Russia,* by M. A. Besborodov, which provides the first chemical analyses of glass of Middle Asia (65). Mention should also be made of the papers of M. Amidjanova (1, 60, 62), I. Akhrarov (63, 64), E. Guliamova (74, 75), E. A. Davidovich (76–80), L. Meregin (101), and B. J. Stavisky (108, 109) who was especially concerned with the glass of Middle Asia.

In 1958, in the discussion of the report on the first "Congrès des Journées internationales du verre," held at Liège, Belgium, Ray W. Smith expressed the opinion that even in the earliest times all large towns of central Asia had a highly developed glass-manufacturing industry. As an example he cited Samarkand and Bukhara. In noting the particular role of Middle Asia in the history of glass manufacturing, he commented: "I think that the products of a glass industry in central Asia, which we have thus far not been able to recognize, will sooner or later prove to have been the bridge between the outright Western production and the Far Eastern production" (129).

There is at present information about the findings of glass productions in many sections in different areas of Middle Asia beginning with the thirteenth through the eleventh centuries B.C. until the fifteenth through seventeenth centuries A.D. In a series of settlements, glass-making shops have been discovered.

This paper describes only the glass products of the pre-Mongolian period. All available materials show a

of glass vessels of the eleventh to the twelfth centuries was discovered.

2. *Medieval Termez* (the right-hand bank of Amu-Darya) in south Uzbekistan. Among various scattered glass objects here, those of the greatest interest are the lockets discovered in the palace. They are made of green and red glass, ornamented with pictures of animals, birds, a woman, and a horseman. On some there are Arabic inscriptions (82, 83, 97, 98, 111).

3. *South Tadzhikistan.* On the site Shakhr-i-Mingh in the plain of the river Vakhsh, fragments were discovered of glass vessels, bracelets, beads, and also lockets with remains of an Arabic inscription (66).

On the site Hisht-Tepe, in the ruins of the town Hulbuk, the capital of the medieval dominion of Huttal, there was found a remarkable collection of glass vessels (goblets, bowls, plates, carafes, etc. (74, 75). Five specimens of glass were submitted to chemical analysis (Table 1).

4. *From Afrasiyab (Samarkand)* fine specimens of various glass products, vessels, window glass and beads were obtained (70, 71, 78, 94, 108, 120). These materials permit one to retrace the whole process of making eyed beads (68). The production of the famous glass blowers from Samarkand was well-known far beyond the limits of Middle Asia (131).

On the site Kuldor-Tepe in the valley of Zarafshan was discovered a glass vessel dating from the second half of the ninth century (109). Five fragments of the vessel from Kuldor-Tepe were subjected to chemical analysis (Table 1).

5. *In the Bukhara oasis* the richest materials of glass-making were discovered on the site Varakhsha where

there probably also existed a glass-making shop (89, 104). Other related findings were made (84, 86–88, 117–119).

6. *In the Tashkent oasis* and on the territory of the town Tashkent several places of findings of medieval glass products are known (70, 99, 113, 118).

7. *Ferghana.* Here on many sites have been found medieval glass products (73, 93, 102).

In the ruins of the medieval Akhsikath (the right-hand bank of Syr-Darya in north Ferghana), which was the capital of Ferghana in the tenth to eleventh centuries, diverse glass vessels have been found (70).

A rich collection of glass has been found in the excavations of the site Kuva in South Ferghana. Here also have been discovered remains of glass furnaces with pieces of glass melt, unfinished goods, and flawed articles. The set of vessels include goblets, mugs, carafes, bottles and ink-stands. Window glass was also found (61, 63, 64).

Specimens of Kuva ware such as jugs, glasses, goblets, wine-glasses, bottles, jars, bracelets and window glass have been subjected to chemical analysis (Table 1).

The medieval Uzgand (right-hand bank of Qara-Darya in east Ferghana), which had been the capital of Ferghana in the eleventh to twelfth centuries, yielded a considerable quantity of glass products (65, 70, 85). Three analyses are given in Table 1.

Various glass vessels also have been found on the site Munchak-Tepe in west Ferghana (72) and on the site Kalai-Bolo in Isfara (77, 79, 80).

8. *Khorezm.* Finds of glass vessels and window glass have been made on the site Shakh-Senem ninety kilometers southwestward from the town Kunia-Urgench. In the neighborhood of the site and within its natural boundaries

Khuyu-Sayi relics of several glass-making shops were found (115). The analyses of glass samples from Khuyu-Sayi, Shakh-Senem and Teke-Senghyr are given further on. The findings in other settlements are well known (69, 116).

9. *South Kazakhstan and the valley of Chu in the North Kirgiz Republic.* Medieval glass has been found on the sites Sairam (96), Krasnaia Rechka (67), Ak Beshim (95). Especially interesting and rich materials were uncovered in the excavations of the site Taraz in the town Djambul (105) where, probably, there had existed glass-making shops. Apart from the places mentioned finds of glass are recorded at other sites (81, 90–92, 100, 103, 106, 107, 110, 112, 116).

## Chemical Investigation of Glass from Middle Asia

In recent years, the necessity for the use of different methods for the study of materials found in excavations, and glass particularly, has been repeatedly noted (65, 124, 127, 130, 133–134).

At present, among several methods of investigation of ancient and medieval glass, special importance is attached to the complete chemical analysis—the only method revealing the composition of the glass and giving the exact quantitative relationship of the components. In 1953, in his report for the Third International Congress on Glass in Venice, after a special discussion of the importance of chemical analysis for the study of ancient glasses, W. E. S. Turner said: "It will be noted that in all four problems chosen for illustration, chemical analysis was the tool employed to provide information from which to draw conclusions." (134.) Later on, he returned to this question and once more pointed out the advantages of chemical analysis applied to the investigation of ancient glass (135, 136).

As mentioned before, some specimens of glass from Middle Asia have been submitted to chemical analysis. Thirty-five speciments from the eighth to the thirteenth centuries A.D., from Piandzhikent, Kuldor-Tepe, Hulbuk, Kuva, Uzgand and Khorezm, have been selected for chemical analysis (56). They are, for the most part, fragments of different vessels. Among those analyzed were vases, jugs, wine glasses, cups, jars and bottles. They were almost colorless or with a faint light-green, light-yellow or light-violet shade. Black or dark-green specimens were rare. Besides these objects a bracelet and a fragment of a flat window-glass were analyzed. All the objects were transparent except a few. A considerable part of the specimens had an iridescent surface. Only the interiors of the specimens, undamaged by weathering, were used for the analysis. The analyses were made according to international standards of practice (126).

In Table 1 are given the average chemical compositions of glass from Middle Asia calculated on the base of chemical analyses for each monument separately (56). In column 8 are given the limit values of oxides in the analyses of all glasses. On the lower horizontal line is given the quantity of analyses for each site separately and of all of them together. In column 9 are given the resulting data—"the average composition of glasses of Middle Asia"—calculated from 22 analyses. The Kuva glass was excluded from the calculations. In the horizontal line after the alkaline oxides are given the totals of the alkalies—potassium oxides and sodium oxides. Attention is called to the limit values of alkaline oxides in column 8. Such great variations of them can be explained by the fact that among the analyzed specimens there prevailed those containing sodium oxide, and several were glass from Kuva in which, on the contrary, potassium oxide prevailed.

The comparison of compositions of the different glass products of the eighth to the thirteenth centuries in Table 1 shows that most of them are very similar chemically and, in fact, belong to the same chemical type. The main mass of the vitreous substance in them consists of the same components. If the glass components depending on their quantitative content can be divided into the *main,* the *secondary* or the *minor* and the *microelements,* then under the main mass of the vitreous substance is to be grouped the total of the main glass components. Some years ago, one of the authors introduced this concept with the purpose of distinguishing in glass the principal chemical characteristics which are important for their chemicotechnological, and other, generalizations (65).

On the basis of Table 1, it can be assumed that in medieval glass production "the main mass of the vitreous substance"—or conventionally "vitreous base"—consists of silica, calcium oxide, magnesium oxide, potassium oxide and sodium oxide. The other oxides occupy a subordinate position and belong to a series of secondary or minor components playing the part of admixtures (aluminium oxide, ferric oxide, manganese oxide, sulfur trioxide).

The analyses of medieval glass which we possess now permit one to reach certain general conclusions on their chemical type. However, so far it is difficult to speak about particularities of separate series, as the number of analyses on each of them is evidently insufficient. Yet some remarks can be made already. Beginning with Kuva glass, the chemical analyses have shown that the thirteen specimens studied can be subdivided into three distinctly differing groups. All the glasses from Kuva are similar among themselves in all their components except the alkaline oxides. The majority of specimens, for which in Table 1 the average values are given, contains in equal, or nearly equal, parts sodium and potassium oxides (sodium oxide, 7.12 to 10.84 percent, and potassium oxide 5.64 to 8.85 percent), one sodium glass (15.26 percent sodium oxide and 0.42 percent potassium oxide) and three potassium glasses (15.28 to 16.43 percent potassium oxide and 0.12 to 0.45 percent sodium oxide). At present it is difficult to explain such a variation of glass compositions within the limits of one site owing to the lack of available materials.

Possibly the relation of potassium and sodium changed in them because of the composition of ash, which was used for the glass-making, for the content of these components is also very alterable. Further studies of Kuva glass and other ones from Middle Asia will permit more precise answers to these questions.

It is interesting to compare glasses from Kuva and from nearby Uzgand in the south and east of Ferghana, respectively. The glasses from Uzgand are distinguished by a raised alkali content compared with those from Kuva. Moreover, the Uzgand glasses contain a lesser quantity of alkaline earth components (8.27 percent and 11.44 percent) compared to the Kuva glasses.

Interest is centered on Khorezm specimens because of their somewhat higher content of silica and a reduced amount of alkali, although these factors are only slight and possibly will not be confirmed by further investigations.

Interesting too, is the comparison of chemical compositions of glasses from Middle Asia and glasses from other countries of the ancient world and of the Middle Ages given in Table 2. For clearer distinctions the table does not give complete chemical analyses, but lists only the contents of five components of the glasses, which among the medieval specimens constitute the "vitreous base." In the calculations of the analyses, their sum appears as 100 percent. In column 6 is given the average composition of ancient glasses of the same type, calculated from the data of columns 2 to 5. Evidently glasses of Middle Asia, in general, are chemically similar to the ancient and medieval specimens of the same type, although they also have some particular attributes; in some a raised content of magnesium oxide and potassium oxide is observed. They differ sharply from ancient Russian lead glasses (column 7 and 8), (65, 125). The known, typical, ancient Russian glasses are of the potassium-lead-silica type. In the territory of Middle Asia to the present time, lead glass has not been discovered. Georgian glasses, made from the fourth to the seventh centuries in ancient Mzkhetha, differ in the high content of manganic oxide from 10.04 to 17.80 percent and have no analogy with any of the medieval glasses (128). The same sort of variation is exhibited by the last of the glasses from Bailakan (medieval Azerbaijan) which have a higher content of alumina (7.24 to 11.22 percent) (123).

The resemblance between the chemical compositions of medieval glasses of Middle Asia and the glasses from other countries of the ancient world (Egypt, Assyro-Babylon, Rome, etc.) permit one to maintain that the craftsmen of Middle Asia used the same formula for the batch as the glass-makers of other countries of those times, i.e., they used one part of sand and three parts of ash (or alkali-salts). There is neither reason nor need to suppose that the craftsmen of Middle Asia introduced into the batch some sort of other components, for ashes of plants of Middle Asia together with sand completely provided the possibility of obtaining glass, the compositions of which are shown in Tables 1 and 2. Indeed the question of the batch for some glasses of Kuva is still under study. The higher content of magnesium oxide and potassium oxide in glasses of Middle Asia compared with some others can be explained by the composition of ashes from plants and, obviously, by the soil of Middle Asia (121).

It is known, that the chemical composition of ancient and medieval glass depends to a considerable degree on the chemical composition of ashes used for their melt (122). In its turn the composition of ashes depended upon the elementary composition of the soil where the plants whose ash was used had grown. In different "biogeochemical provinces," the soils differ in the content of mineral components and, particularly in microelements.

In general one could say that the composition of ancient glass bears the imprint of the chemical composition of the soil where it was made. For a successful development of the archeological technology of glass and to find the local particularities of ancient and medieval glasses, investigators in the future will be obliged to take note not only of the complete chemical analyses of the glasses, but also of the chemical composition of the soil from which the ash derives.

If today science considered data concerning soils on the territories of ancient and medieval civilizations, the question, for example, of the origin of the Afghanistani glass discovered in the cave Shamshir-Ghar and containing such a rare microelement as rubidium should be easy to solve (131). The rubidium in this glass is probably a local characteristic.

There are sound reasons to affirm that the second component, besides the sand the workmen of Middle Asia introduced into the batch, was the *Ishkor* (65), as it is called in Middle Asia, i.e., the ashes of alkali-containing plants representing a sintered, somewhat vitrified mass of a spongy structure and a light-grey or green-grey color. *Ishkor* consists mainly of a mixture of carbonates, sulphates and chlorides of sodium, potassium, calcium and magnesium; in small quantities silica, alumina, ferric oxide also are present (121). To obtain *Ishkor,* plants belonging mostly to the family of *Chenopodiaceae* are used. It is known that in Assyro-Babylon these plants were used to obtain the ash (132).

Very likely the glass-makers of Middle Asia used also the ashes of cane. In Middle Asia several alkali-bearing plants grow which could completely supply the local glass-making shops with ash. No import of alkaline raw materials was required.

All materials accessible at present show a considerably widespread use of glassware in antiquity and the Middle Ages in Middle Asia. The origin of local glass-making probably is in the last centuries B.C., but the assortment of glass products in the ancient period was still very restricted.

The glass-making craft attained a wide development in the Middle Ages, when probably each fairly large town or settlement had its own glass-making shop. In some cases this is confirmed by finds of remains of glass-making shops, in others by various indirect proofs. At that time there was a great diversity of functional ware: dishes, adornments, everyday objects, technical and medical accessories. Window glass was also widespread. The study of archeological materials permits one to consider Middle Asia as one of the centers of glass-making in the Middle Ages. Of interest

TABLE I.

Chemical Composition of Medieval Glass
8th–13th Centuries A.D. in %

| *Oxides* | *Piandzhikent 8th cent.* | *Kuldor-Tepe 9th cent.* | *Hulbuk 10th–12th cent.* | *Uzgand 11th–13th cent.* | *Khorezm 12th–14th cent.* | *Kuva 10th–12th cent.* | *Extreme Values of Oxides in Glass Analyses* | *Average Compositions (without Kuva)* |
|---|---|---|---|---|---|---|---|---|
| 1 | 2 | 3 | 4 | 5 | 6 | 7 | 8 | 9 |
| $SiO_2$ | 65,35 | 64,63 | 67,47 | 67,45 | 68,44 | 66,66 | 62,87 – 70,04 | 66,67 |
| $Al_2O_3$ | 2,10 | 2,43 | 1,48 | 1,98 | 1,95 | 2,92 | 0,97 – 3,45 | 1,99 |
| $Fe_2O_3$ | 1,05 | 0,79 | 0,83 | 0,30 | 1,18 | 1,80 | 0,21 – 2,35 | 0,83 |
| CaO | 7,76 | 7,48 | 7,52 | 5,62 | 6,76 | 6,72 | 4,73 – 8,68 | 7,28 |
| MgO | 4,30 | 4,63 | 4,85 | 2,65 | 4,84 | 4,72 | 2,62 – 6,02 | 4,25 |
| $Mn_2O_3$ | 0,32 | 0,88 | 0,56 | 1,83 | 0,08 | 0,40 | 0 – 2,47 | 0,73 |
| $SO_3$ | 0.15 | 0,49 | 0,06 | 0,84 | 0,17 | 0,33 | 0,02 – 0,96 | 0,34 |
| $K_2O$ | 3,77 | 3,70 | 2,42 | 2,40 | 3,94 | 7,15 | 0,42 – 16,43 | 3,25 |
| $Na_2O$ | 15,12 | 14,92 | 14,89 | 16,66 | 12,85 | 8,92 | 0,12 – 17,54 | 14,89 |
| $K_2O \div Na_2O$ | 18,89 | 18,62 | 17,31 | 19,06 | 16,79 | 16,07 | – | 18,13 |
| *Number of analyses* | 4 | 5 | 5 | 3 | 5 | 9 | 35 | 22 |

TABLE II.

Chemical Composition of Ancient and Medieval Glass in %:
The Main Components

| | | | | | | Ancient Russia | | Middle Asian | |
|---|---|---|---|---|---|---|---|---|---|
| *Oxides* | *Egyptian* | *Assyro-Babylonian* | *Indian* | *Roman* | *Average Content* | *Lead oxide-Silica* | *Potassium oxide-Lead oxide-Silica* | *All Except Glasses from Kuva* | *Kuva (sodium-potassium glasses)* |
| 1 | 2 | 3 | 4 | 5 | 6 | 7 | 8 | 9 | 10 |
| $SiO_2$ | 69,1 | 71,2 | 70,3 | 71,0 | 70,3 | 27 | 57 | 70,56 | 70,86 |
| CaO | 7,5 | 6,8 | 5,5 | 7,3 | 7,2 | – | – | 7,42 | 7,14 |
| MgO | 3,1 | 4,9 | 2,7 | 1,1 | 2,6 | – | – | 4,51 | 5,02 |
| PbO | – | – | – | – | – | 73 | 28 | – | – |
| $K_2O$ | 2,1 | 2,4 | 3,4 | 1,2 | 2,0 | – | 15 | 3,37 | 7,61 |
| $Na_2O$ | 18,2 | 14,7 | 18,1 | 19,4 | 19,9 | – | 15 | 19,13 | 17,09 |
| $K_2O \div Na_2O$ | 20,3 | 17,1 | 21,5 | 20,6 | 17,9 | – | – | 15,76 | 9,48 |
| *Number of analyses* | 71 | 18 | 24 | 55 | 168 | 13 | 33 | 22 | 9 |

Note: The data of the compositions in columns 2–8 is taken from the monograph by M. A. Besborodov: "Glassmaking in Ancient Russia," Minsk, 1956.

is the fact that, according to the evidence, glass goods were not imported. To the contrary, it is known that some kinds of glassware were exported from Middle Asia, particularly from Khorezm to Voljskie, Bulgaria (115).

All available materials, in sum, show the high development of the glass-making craft in the pre-Mongolian time in Middle Asia.

The Mongol invasion caused an enormous damage to glass-making and arrested its development. In the thirteenth to the sixteenth centuries, glassware is encountered in reduced quantity and at that time glass had not its former significance. However one cannot speak of a complete suspension of glass-making. It continued to exist, for instance in the Khorezm.

## ЛИТЕРАТУРА

### Древний период

1. М. Аминджанова. *Два сосуда из египетского стекла в собрании Самаркандского музея.* Научные работы и сообщения ООН АН УзССР, 2, 1961.
2. Г. Г. Бабанская. *Берккаринский могильник.* ТИИАЭ АН КазССР, 1, 1956.
3. В. В. Бартольд. *История культурной жизни Туркестана.* Л., 1927.
4. Ю. Д. Баруздин. *Кара-Булакский могильник.* (Раскопки 1954 г.), ТИИ АН КирССР, II, 1956.
5. А. Н. Бернштам. *Кенкольский могильник.* Л., 1940.
6. ———*Из итогов археологических работ на Тянь-Шане и Памиро-Алае.* КСИИМК, 28, 1949.
7. ———*Историко-археологические очерки Центрального Тянь-Шаня и Памиро-Алая.* МИА № 26, 1952.

8. Н. Я. Бичурин. *Собрание сведений о народах, обитавших в Средней Азии в древние времена.* М.-Л., 1950.
9. Н. И. Веселовский. *Заметка о стеклянном производстве в Средней Азии.* Записки Вост. Отделения Русского Археологического Общества, VIII, 1894.
10. М. В. Воеводский и М. П. Грязнов. *У-суньские могильники на территории Киргизской ССР,* ВДИ, 1938. № 3-4.
11. М. Э. Воронец. *Археологические исследования Института истории и археологии и Музея истории Академии Наук УзССР на территории Ферганы в 1950-1951 годах.* Труды Музея истории УзССР, II, 1954.
12. В. Ф. Гайдукевич. *Могильник близ Ширин-Сая в Узбекистане.* СА, XVI, 1952.
13. Б. З. Гамбург и Н. Г. Горбунова. *Могильник эпохи бронзы в Ферганской долине.* КСИИМК, 63, 1956.
14. ———*Ак-Тамский могильник.* КСИИМК, 69, 1957.
15. ———*Новые данные о культуре эпохи бронзы Ферганской долины.* СА, № 3, 1957.
16. ———*Археологические работы Ферганского областного краеведческого музея в 1953-1954 гг.* (краткий отчет). История материальной культуры Узбекистана. Вып. 1, 1959.
17. ———*Могильник Хангиз.* ИООН АН ТаджССР, 14, 1957.
18. М. М. Дьяконов. *Работы Кафирниганского отряда.* МИА № 15, 1950.
19. В. Д. Жуков. *Материалы к изучению Баш-тепинской группы памятников в западной части Бухарского оазиса.* ТИИА АН УзССР, VIII, 1956.
20. Ю. А. Заднепровский. *Археологические памятники южных районов Ошской области.* Фрунзе, 1960.
21. М. А. Итина. *Раскопки могильника тазабагъябской культуры Кокча 3.* Материалы Хорезмской экспедиции, 5, 1961.

22. А. Кибиров. *Археологические памятники Чаткала.* ТКАЭЭ, II, 1959.
23. П. И. Князев. *Разведочно-археологические работы в квартале металлистов древнего Термеза.* ТАН УзССР. ТАКЭ II, 1945.
24. И. Кожомбердиев. *Могильник Акчий-Карасу в долине Кетмень-Тюбе.* ИАН КирССР. Серия обществ. наук, т. II, вып. 3, 1960.
25. Б. А. Литвинский. *Об изучении в 1955 г. погребальных памятников кочевников в Кара-Мазарских горах.* ТАН ТаджССР, 63, 1956.
26. Б. А. Литвинский. *Предварительный отчет о работах в Кара-Мазарских горах отряда по сбору материалов для составления археологической карты в 1954 г.* ТАН ТаджССР, 37, 1956.
27. ———*Предварительный отчет о раскопках курганов в Ворухе* (Исфаринский район) *в 1954 г.* ТАН ТаджССР, 37, 1956.
28. ———*Изучение курумов в северо-восточной части Ленинабадской области.* ТАН ТаджССР, 102, 1959.
29. ———*Раскопки могильников в Исфаринском районе в 1956 г.* ТАН ТаджССР, 91, 1959.
30. ———*Исследование могильников Исфаринского района в 1958 г.* ТИИАН ТаджССР, XXVII, 1961.
31. А. Г. Максимова. *Могильник эпохи бронзы в урочище Тау-Тары.* ТИИАЭ АН КазССР, 14, 1962.
32. Л. Я. Маловицкая. *Тамдинский курганный могильник III-I вв. до н. э.* ИАН КазССР, серия археолог., 2, 1949.
33. А. М. Мандельштам. *Археологические работы в Бишкентской долине в 1957 г.* ТАН ТаджССР, 103, 1959.
34. ———*Новые данные о Тулхарском могильнике по работам 1958 г.* ТИИАН ТаджССР, XXVII, 1961.

35. А. А. Марущенко. *Хосров-Кала* (Отчет о раскопках 1953 г.). ТИИАЭ АН ТуркССР, II, 1956.
36. М. Е. Массон. *Народы и области южной части Туркменистана в составе Парфянского государства.* ТЮТАКЭ, V, 1955.
37. М. Е. Массон и Г. А. Пугаченкова. *Парфянские ритоны Нисы.* ТЮТАКЭ, IV, 1959.
38. Е. Е. Неразик. *Археологическое обследование городища Куня-Уаз в 1952 г.* ТХАЭЭ, II, 1958.
39. О. В. Обельченко. *Курганные погребения первых веков н. э. и кенотафы Кую-Мазарского могильника* ТСАГУ, CXI, Историч. науки, 25, 1957.
40. ———*Лявандакский могильник.* История материальной культуры Узбекистана. 2, 1961.
41. И. В. Пташникова. *Бусы древнего и раннесредневекового Хорезма.* ТХАЭЭ, 1, 1952.
42. С. С. Сорокин. *Боркорбазский могильник* (Южная Фергана, бассейн реки Сох). Труды Гос. Эрмитажа, V, 1961.
43. А. И. Тереножкин. *Согд и Чач.* КСИИМК, XXXIII, 1950.
44. С. П. Толстов. *Древний Хорезм.* М., 1948.
45. К. В. Тревер. *Памятники греко-бактрийского искусства.* Л., 1940.
46. С. А. Трудновская. *Украшения позднеантичного Хорезма по материалам раскопок Топрак-Кала.* ТХАЭЭ, 1, 1952.

## Раннесредневековый период

47. Л. И. Альбаум. *Балалык-Тепе.* Ташкент, 1960.
48. А. М. Беленицкий. *Общие результаты раскопок городища Древнего Пенджикента (1951-1953 гг.).* МИА № 66, 1958.

49. И. Б. Бентович. *Находки на горе Муг.* МИА № 66, 1958.
50. А. Н. Бернштам. *Труды Семиреченской археологической экспедиции. Чуйская долина.* МИА № 14, 1950.
51. О. Г. Большаков и Н. Н. Негматов. *Раскопки в пригороде Древнего Пенджикента.* МИА № 66, 1958.
52. С. А. Ершов. *Некоторые итоги археологического изучения некрополя с оссуарными захоронениями в районе города Байрам-Али.* ТИИАЭ АН ТуркССР, V, 1959.
53. А. М. Мандельштам и С. Б. Певзнер. *Работы Кафирниганского отряда в 1952-1953 гг.* МИА № 66, 1958.
54. В. М. Массон. *К истории парфянского и раннесредневекового Дахистана.* ИАН ТуркССР, Серия обществ. наук, 1961, № 2.
55. И. В. Пташникова. *Ук. раб. № 41.*

## Средневековый период

56. А. Абдуразаков и М. А. Безбородов. *Химическое исследование средневековых стекол Средней Азии.* Узбекский химический журнал, 1962, № 3.
57. К. Адыков. *К характеристике гончарного производства в Мерве, конца XII — начала XIII вв.* ИАН ТуркССР, 1955, № 6.
58. ———*Главные станции на средневековом торговом пути из Серахса к Мерв.* СА, 1959, № 4.
59. ———*Тильситана.* ИАН ТуркССР, 1959, № 1.
60. М. Аминджанова. *О некоторых сосудах Мавераннахра.* История материальной культуры Узбекистана. 2, 1961.
61. ———*О производстве стеклянных изделий в средневековом городе Кува.* Научные работы и сообщ. ООН. АН УзССР, I, 1960.
62. ———*Стеклянный прибор для умывания в собрании*

*Ташкентского музея.* Научные работы и сообщ. ООН АН УзССР. IV, 1961.

63. Ахраров, И. *Средневековые стеклянные бокалы из Кувы.* ИАН УзССР, Серия обществ. наук, 1960, № 4.
64. ———*Средневековые чернильницы с городища Кува.* Обществен. науки в Узбекистане. 1962, № 1.
65. М. А. Безбородов. *Стеклоделие в Дравней Руси.* Минск, 1956.
66. А. М. Беленицкий. *Отчет о работе Вахшского отряда в 1946 г.* МИА № 15, 1950.
67. А. Н. Бернштам. *Труды Семиреченской археологической экспедиции. Чуйская долина.* МИА № 14, 1950.
68. А. С. Боброва. *Бусы из Афрасиаба. Следы производства глазчатых стеклянных бус в Средней Азии.* КСИИМК, XXX, 1949.
69. Н. Н. Вактурская. *О раскопках 1948 г. на средневековом городе Шемаха-Кала Туркменской ССР.* ТхаЭЭ I, 1952.
70. Н. И. Веселовский. (сообщение о раскопках в Туркестанском крае). *Отчет Археологической Комиссии 1882-1888.* СПб, 1891.
71. В. Л. Вяткин. *Афросиаб — городище былого Самарканда.* 1926.
72. В. Ф. Гайдукевич. *Работы Фархадской археологической экспедиции в Узбекистане 1943-1944 гг.* КСИИМК, XIV, 1947.
73. Я. Г. Гулямов. *Отчет о работе третьего отряда археологической экспедиции на строительстве Большого Ферганского канала.* ТИИА АН УзССР, IV, 1951.
74. Э. Гулямова. *Раскопки цитадели на городище Хульбук в 1957 г.* ТИИ АН ТаджССР, XXVII, 1961.
75. ———*Стекло с городища Хульбук.* ИООН АН ТаджССР, I (24), 1961.

76. Е. А. Давидович. *Стекло из Нисы.* ТЮТАКЭ, I, 1949.
77. ———*Средневековое оконное стекло из Таджикистана.* Доклады АН ТаджССР, 7, 1953.
78. ———*Цветное оконное стекло XV в. из Самарканда.* ТСАГУ, 61, 1953.
79. ———*Раскопки замка Калаи-Боло.* МИА № 66, 1958.
80. Е. А. Давидович и Б. А. Литвинский. *Археологический очерк Исфаринского района.* ТАН ТаджССР, 35, 1955.
81. С. А. Ершов. *Данданакан.* КСИИМК, XV, 1947.
82. В. Д. Жуков. *Стеклянные "медальоны" из дворца термезских правителей.* Изв. УзФАН СССР, 1940, 4-5.
83. ———*Археологическое обследование в 1937 г. дворца термезских правителей.* ТАН УзССР. ТАКЭ, II, 1945.
84. ———*Археологическая разведка на шахристане Хайрабад-Тепе.* История материальной культуры Узбекистана, 2, 1961.
85. Ю. А. Заднепровский. *Археологические работы в Южной Киргизии в 1954 году.* ТКАЭЭ, IV, 1960.
86. Л. А. Зимин. *Развалины Старого Пейкенда.* ПТКЛА, XVIII, 1913.
87. ———*Отчет о весенних раскопках в развалинах Старого Пейкенда.* ПТКЛА, XIX, 1915.
88. ———*Отчет о летних раскопках в развалинах Старого Пейкенда.* ПТКЛА, XIX 1915.
89. С. К. Кабанов. *Раскопки жилого квартала X века в западной части городища Варахша.* ТИИА АН УзССР, VIII, 1956.
90. И. А. Кастанье. *Древности Ура-Тюбе и Шахристана.* ПТКЛА, XX, 1915.
91. А. Кибиров. *ук. раб.*

92. А. К. Кларе. *Древний Отрар и раскопки, произведенные в развалинах его в 1904 году.* ПТКЛА, IX, 1904.
93. В. И. Козенкова. *Археологические работы в Андижанской области в 1956 г.* КСИИМК, 76, 1959.
94. С. А. Кудрина. *Анализ цветного стекла из мавзолея Ишрат-Хана. Сб. Мавзолей Ишрат-Хана.* Ташкент. 1958.
95. Л. Р. Кызласов. *Археологические исследования на городище Ак-Бешим в 1953-1954 гг.* ТКАЭЭ, II, 1959.
96. М. Е. Массон. *Старый Сайрам.* Известия Средазкомстариса. III, 1928.
97. ———*Городище Старого Термеза и их изучение.* Труды УзФАН СССР, ТАКЭ. 1940.
98. ———*Работы Термезской археологической комплексной экспедиции (ТАКЭ) 1937 и 1938 гг.* ТАН УзССР, ТАКЭ, II, 1945.
99. ———*Ахангеран. Археолого-топографический очерк.* Ташкент. 1953.
100. ———*Обнаружение и опознание руин средневекового селения Шавваль.* ИАН ТуркССР, 6, 1959.
101. Л. Мережин. *К характеристике средневекового стекла из Мерва.* Сборник студенческих работ САГУ, XV, 1956.
102. Н. Негматов. *Предварительный отчет о работах Ходжентского отряда в 1954 г.* ТАН ТаджССР, XXXVII, 1956.
103. Н. Негматов, Т. И. Зеймаль. *Раскопки на Тирмизак-Тепе.* ИООН АН ТаджССР, I (24), 1961.
104. В. А. Нильсен. *Варахшская цитадель.* ТИИА АН УзССР, VIII, 1956.
105. Г. И. Пацевич. *Раскопки на территории древнего города Тараза в 1940 году.* ТИИАЭ КазССР, I, 1956.
106. А. А. Росляков. *Мелкие археологические памятники окрестностей Ашхабада.* ТЮТАКЭ, V, 1955.

107. Б. Я. Ставиский. *Археологические работы в бассейне Магиан-Дарьи в 1957 г.* ТАН ТаджССР, СШ, 1952.

108. ———*Самаркандские чернильные приборы IX-X вв. в собрании Эрмитажа.* СА, 1960, № 1.

109. Б. Я. Ставиский. *Раскопки городища Кулдор-Тепе в 1956-1957 гг.* СА, 1960, № 4.

110. ———*Археологические работы в районе Магиан-Дарьи в 1957-1959 гг.* Сообщения Гос. Эрмитажа, XXI, 1961.

111. И. А. Сухарев. *Комната XII века на площади Приамударьинской части городища Термеза.* ТАН УзССР. ТАКЭ, 1940.

112. А. И. Тереножкин. *Археологическая рекогносцировка в западной части Узбекистана.* ВДИ, 1947, № 2.

113. ———*Холм Ак-Тепе близ Ташкента* (Раскопки 1940 г.). ТИИА АН УзССР, I, 1948.

114. ———*ук. раб. № 43.*

115. С. А. Трудновская. *Стекло с городища Шах-Сенем.* ТХАЭЭ, II, 1958.

116. Г. А. Федоров-Давыдов. *Раскопки квартала XV-XVII вв. на городище Таш-Кала.* ТХАЭЭ, II, 1958.

117. В. А. Шишкин. *Мечеть Магаки-Аттари в Бухаре.* ТИИА АН УзССР, I, 1948.

118. ———*Узбекистанская археологическая экспедиция АН УзССР* (полевые работы 1956-1959 гг.) История материальной культуры Узбекистана, 2, 1961.

119. Г. В. Шишкина. *Археологическая разведка на городище Кумсултан* (Бухарская область). *Научные работы и сообщ.* ООН АН УзССР, 3, 1961.

120. А. Ю. Якубовский. *Среднеазитские собрания Эрмитажа и их значение для изучения истории культуры и искусства Средней Азии до XVI в.* Труды Отдела Востока Эрмитажа, II, 1940.

## Химическое исследование

121. А. Абдуразаков. *Сырьевые ресурсы раннесредневекового стеклоделия Средней Азии.* Сборник Трудов молодых химиков УзбССР, Ташкент, 1962 (в печати).

122. М. А. Безбородов. *Древнерусские стекла и огнеупорные изделия.* КСИИМК, 62, 1956, он же, Стеклоделие в древней Руси, стр. 211.

123. М. А. Безбородов и А. Л. Якобсон. *Химическое исследование средневековых стекол из Байлакана.* СА, 1960, № 4.

124. M. A. Besborodow. *Glasherstellung bei den slawischen Völkern an der Schwelle des Mittelalters.* Wissenschaftliche Zts. der Humbolt-Universität zu Berlin, 1958/1959. Jg. VIII, Nr. 2/3. S. 187-193.

125. *"A Chemical and Technological Study of Ancient Russian Glasses and Refractories."* Journ. Soc. Gl. Techn., 1957, vol. XLI, pp. 168-184.

126. В. Ф. Гиллебранд и Г. Э. Ленделль. *Практическое руководство по неорганическому анализу.* Москва, 1935.

127. Б. А. Колчин и А. Л. Монгайт. *Археология и методы естественных наук.* Вестник АН СССР, 1959, № 12.

128. Н. Н. Угрелидзе. *К истории производства стекла в древней Грузии* (Стеклянные сосуды IV-VIII вв. н. э. из Самтаврского могильника), Тбилиси, 1955.

129. *Annales du Ier Congrès des "Journées internationales du Verre".* (Liège, 1958), p. 42.

130. R. H. Brill and H. P. Hood. *"A New Method for Dating Ancient Glass." Nature,* Vol. 189, No. 4758, 1961, pp. 12-14.

131. Louis Dupree. *Shamshir Ghar: "Historic Cave Site in Kandahaz Province, Afghanistan, Anthropological Papers of the American Museum of Natural History,* Vol. 46,

part 2 (New York, 1958). Технический доклад W. R. Smith об образцах стекла из Шамшир Гар, стр. 289)

132. M. Levey. *Chemistry and Chemical Technology in Ancient Mesopotamia.* (Amsterdam, 1959).

133. R. W. Smith. *"Technological Research on Ancient Glass."* Archeology, 1958, Vol. 11, No. 2, pp. 111-116.

134. W. E. S. Turner. *"The Value of Modern Technical Methods in the Study of Ancients Glasses."* Third International Congress on Glass, Venice, 1953.

135. ———*"The Analysis of Ancient Glasses."* The Glass Industry, 1955, No. 7.

136. ———*"Studies in Ancient Glasses and Glassmaking Processes."* Part IV, *The Chemical Composition of Ancient Glasses,"* Journ. Soc. Gl. Techn., Vol. XL, No. 193 (1956), pp. 162-186.

## Список сокращений

ВДИ — Вестник Древней Истории.

ИАН — Известия Академии Наук.

Изв. УзФАН — Известия Узебкского Филиала Академии Наук СССР.

ИООН — Известия Отделения Общественных Наук.

КСИИМК — Краткие Сообщения Института Истории Материальной Культуры АН СССР.

МИА — Материалы и исследования по археологии СССР

ООН — Отделение Общественных Наук.

ПТКЛА — Протоколы Туркестанского Кружка Любителей Археологии.

СА — Советская Археология.

ТАКЭ — Термезская археологическая комплексная экспедиция.

ТАН — Труды Академии Наук.

ТИИАН — Труды Института Истории Академии Наук.

ТИИА АН — Труды Института Истории и археологии Академии Наук.

ТИИАЭ АН — Труды Института Истории, археологии и этнографии Академии Наук.

ТКАЭЭ — Труды Киргизской археолого-этнографической экспедиции.

ТСАГУ — Труды Среднеазиатского Гос. Университета.

ТХАЭЭ — Труды Хорезмской археолого-этнографической экспедиции.

ТЮТАКЭ — Труды Южно-Туркменистанской археологической комплексной экспедиции.

# Investigations on the Origin and Manufacture of Orichalcum

By Earle R. Caley
Ohio State University

The Greek name *oreichalkos* and the corresponding Latin name *orichalcum*, otherwise spelled *aurichalcum*, designated different metals at different times,[1] but this Latin name and its equivalent in modern languages is now used, chiefly by numismatists, to designate a particular copper alloy containing zinc which the Romans employed in very late republican times and in imperial times as a material for coins. Some writers on Roman numismatics imply or state that orichalcum was an alloy of fixed composition, but this is not in accord with the facts. At no time were the proportions of copper and zinc held constant, and over the years of its use, there was a marked decrease in the

[1] Rossignol, J. P., *Les métaux dans l'antiquité* (Paris, 1863), pp. 205–331, deals in detail with the etymology of these words and their various meanings. The spelling *orichalcum* is definitely to be preferred to *aurichalcum*. For additional etymological information see: Diergart, P., *Zeitschrift für angewandte Chemie,* XIV (1901), pp. 1297–1301; Ibid., XV (1902), pp. 761–763; Ibid., XVI (1903), pp. 85–88. Disagreement with some of the conclusions of Diergart was expressed by Neumann, B., *Zeitschrift für angewandte Chemie,* XV (1902), pp. 511–516, 1217–1218.

average proportion of zinc, while the proportions of tin and lead, which in the beginning were present in very small amounts as impurities, increased to such an extent that one or the other, or both, became important components of the alloy. These changes in composition are illustrated by the chronological series of analytical data in Table I, data which are representative of a larger number of analyses of orichalcum coins made by various students in the author's laboratory during the past twenty-five years. In a strict sense, the term orichalcum should be understood to refer not to a single alloy but to a class of alloys containing copper and zinc as the principal components.

Although the Romans were undoubtedly the first to employ copper alloys containing zinc as a material for coins, it is not true, as has sometimes been stated, that this was the first use of such alloys for any purpose. The comprehensive researches of Otto and Witter [2] and their coworkers show that copper alloys containing moderate proportions of zinc were occasionally produced even in the Early Bronze Age in Central Europe, and a few prehistoric objects composed of such alloys have been found in Northern Europe and the British Isles. The earliest known metal objects of the Mediterranean region that contain significant proportions of zinc were found at Gezer in Palestine.[3] In most of these the zinc content is less than 4 percent, and could be regarded as a mere accidental impurity. However, one of the objects from this site was found to contain 23.40 percent of zinc, along with 10.17 percent of tin, and 66.40 percent of copper. This object,

[2] Otto, H., and Witter, W., *Handbuch der ältesten vorgeschichtlichen Metallurgie in Mitteleuropa* (Leipzig, 1952), pp. 210–211.

[3] Macalister, R. A. S., *The Excavations of Gezer* (London, 1912), Vol. II, pp. 265, 293, 303. The analyses were made by J. E. Purvis.

TABLE I.

Analyses of Orichalcum Coins of the Roman Empire

| No. | *Emperor* | Cu % | Zn % | Sn % | Pb % | Fe % | Ni % | Ag % | As % | S % | *Others* % | *Total* % |
|---|---|---|---|---|---|---|---|---|---|---|---|---|
| 1 | *Augustus* | 76.70 | 22.02 | 0.27 | 0.37 | 0.49 | 0.04 | 0.04 | tr. | none | none | 99.93 |
| 2 | *Tiberius* | 76.86 | 22.85 | 0.03 | 0.04 | 0.26 | 0.03 | none | none | none | none | 100.07 |
| 3 | *Caligula* | 72.63 | 26.71 | 0.02 | 0.16 | 0.44 | 0.01 | 0.03 | 0.04 | n.d. | none | 100.04 |
| 4 | *Caligula* | 78.19 | 21.11 | 0.12 | 0.04 | 0.39 | none | none | none | none | none | 99.85 |
| 5 | *Claudius* | 75.91 | 23.20 | 0.09 | 0.12 | 0.67 | 0.02 | 0.01 | 0.04 | none | Sb(0.01) | 100.06 |
| 6 | *Claudius* | 77.59 | 21.11 | 0.10 | 0.83 | 0.47 | 0.02 | 0.03 | 0.02 | none | none | 100.17 |
| 7 | *Nero* | 77.27 | 22.46 | none | 0.15 | 0.16 | none | 0.08 | 0.85 | none | none | 100.17 |
| 8 | *Nero* | 83.16 | 15.95 | 0.01 | 0.13 | 0.40 | 0.03 | 0.06 | n.d. | n.d. | n.d. | 99.74 |
| 9 | *Nerva* | 83.60 | 14.82 | 0.70 | 0.54 | 0.46 | 0.03 | none | 0.03 | n.d. | Au(tr.) | 100.18 |
| 10 | *Trajan* | 82.08 | 14.10 | 2.05 | 1.29 | 0.37 | 0.04 | 0.09 | 0.03 | 0.08 | none | 100.13 |
| 11 | *Ant. Pius* | 86.51 | 11.14 | 1.69 | 0.11 | 0.33 | 0.03 | 0.10 | 0.10 | 0.17 | none | 100.18 |
| 12 | *Ant. Pius* | 89.29 | 9.38 | 0.16 | 0.35 | 0.55 | 0.04 | 0.05 | 0.12 | 0.21 | none | 100.15 |
| 13 | *M. Aurelius* | 87.86 | 9.06 | 2.03 | 0.23 | 0.33 | 0.03 | 0.10 | 0.08 | 0.28 | Sb(0.05) | 100.05 |
| 14 | *M. Aurelius* | 88.96 | 7.87 | 2.33 | 0.18 | 0.31 | 0.04 | 0.05 | 0.21 | 0.17 | Sb(0.10) | 100.22 |
| 15 | *Commodus* | 86.85 | 6.43 | 1.64 | 4.28 | 0.28 | 0.06 | 0.07 | 0.13 | 0.32 | none | 100.06 |

the metal of which may be termed zinc bronze, is believed to belong to the period known as Semitic III (1400–1000 B.C.). No other object from this region in which zinc is present with copper as a principal component of an alloy can be dated with certainty prior to the first century B.C. The only metal objects of such composition that may be classified as Greek were found at or near Greek colonies or settlements on the Black Sea. Four such objects are known, all of which were analyzed by Bibra.[4] The results of his analyses are listed in Table II. On the basis of the probable dates of the sites, all four are believed to date from the first century B.C. However, there is a possibility that the actual date of manufacture of some or all of these objects was somewhat earlier than the first century B.C.

The copper alloys which appeared before orichalcum differ in various respects from this Roman alloy, more particularly from the type of orichalcum initially used for coins. Their zinc content varies over a wide range, in the prehistoric alloys especially, and, as is shown in Table III, these earlier alloys contain lower proportions of zinc and higher proportions of metals other than copper and zinc. Groups I and II in this table include all objects for which reliable quantitative data are available and in which the reported zinc content is 5 percent or more. Another important difference is the sporadic occurrence with respect to place and time of the objects composed of these earlier alloys and their great scarcity as compared to objects of similar age composed of tin bronze. In contrast, orichalcum coins were regularly produced in a few localities in enormous numbers for nearly two centuries beginning with the reign of Augustus. All these differences indicate that these

[4] Bibra, E. von, *Die Bronzen und Kupferlegirungen der alten und ältesten Völker* (Erlangen, 1869), pp. 86–87.

earlier alloys were produced accidentally and that orichalcum was manufactured intentionally. A possible exception

TABLE II.

Analyses of Certain Pre-Roman Metal Objects from Sites on the Northern Side of the Black Sea

| *Object No.* | Cu % | Zn % | Sn % | Pb % | Fe % | Ni % | *Total* % |
|---|---|---|---|---|---|---|---|
| 1 | 82.76 | 13.31 | 3.40 | 0.19 | tr. | 0.34 | 100.00 |
| 2 | 84.87 | 10.12 | 4.36 | 0.21 | tr. | 0.44 | 100.00 |
| 3 | 91.00 | 9.00 | none | tr. | tr. | tr. | 100.00 |
| 4 | 90.59 | 8.10 | none | ft. | 1.31 | none | 100.00 |

*Descriptions*

No. 1. Fishhook from a Greek grave on the Crimea.
No. 2. Fishhook from a Greek grave on the Crimea.
No. 3. Ring made of fine twisted wire from the ruins of ancient Tanaïs.
No. 4. Wire from a grave in the vicinity of ancient Olbia.

*Analytical Notes*

a. No. 2 was reported to contain a trace of cobalt and Nos. 3 and 4 each a trace of antimony.
b. Very likely the reported percentages of nickel are too high since it is known that the method used by Bibra for the determination of this element yields high results.
c. The perfect summations are due to Bibra's usual practice of determining copper by difference.

may be some of the copper-zinc alloys produced, or at least found, in the vicinity of the Black Sea, as seems to be indicated especially by the composition of the objects numbered 3 and 4 in Table II. Though these have a lower zinc content, they resemble early orichalcum in their low content of metals other than copper and zinc.

Any inquiry into the origin and technology of orichalcum should include a consideration of the possibility that ancient metallurgists may have isolated metallic zinc, and the further possibility that they may have used it to form alloys with copper. Until recently, clear archeological evidence for the ancient isolation of zinc has not existed, either because reliable chemical tests were not made on the objects alleged to be composed of zinc or because of doubt as to their antiquity. In 1939, one specimen of definitely known provenance, later identified beyond doubt as metallic zinc, was found in the course of the excavation of the Agora at Athens.[5] As to provenance, the discoverer, Dr. Arthur Parsons, reported as follows:

> The fragment of zinc was found in section OA on May 13, 1939, at the bas of the cliff on the north slope of the Acropolis, at a point of about 7.0 meters east of the ancient fountain house, the Klepsydra, and directly below the cave sanctuary of Pan. The pottery and coins found with it were chiefly of the fourth and third centuries B.C.; there was nothing later than the second century B.C. It may be regarded as certain that the zinc got there no later.

An analysis made in the Research Laboratory of the New Jersey Zinc Company showed that it was about 98.7 percent pure and that it contained a greater variety of impurities than are present in modern zinc.

Because of the chemical reactivity of the metal, the rarity of ancient zinc objects might conceivably be ascribed to their general disappearance through corrosion. However, if metallic zinc really had been abundant and in general use in Greek and Roman times, it seems likely that many zinc objects particularly those of thick metal, would have survived in protected locations. The strongest indication that zinc was not abundant or widely available would seem to be the rather low zinc content of the ancient copper-zinc alloys. If metallic zinc had been freely avail-

[5] Farnsworth, M., Smith, C. S., and Rodda, J. L., *Hesperia,* Supplement VII (1949), pp. 126–129.

able for alloying with copper, alloys of higher zinc content would probably have often been manufactured, and a considerable number of objects composed of such alloys would have survived.

TABLE III.

Average Proportions of Copper, Zinc, and Other Metals in Early Copper Alloys Containing Zinc

| *Group No.* | *Objects Analyzed* | *Kind and Source of Objects* | *Cu* % | *Zn* % | *Other Metals* % |
|---|---|---|---|---|---|
| I | 9 | Early Bronze Age Central Europe | 80.7 | 12.4 | 6.5 |
| II | 9 | Prehistoric British Isles and Northern Europe | 79.4 | 16.2 | 4.6 |
| III | 4 | Greek, Northern Black Sea Coast | 87.3 | 10.1 | 2.6 |
| IV | 6 | Orichalcum Coins, Roman Republic and Empire, First Century, B.C. | 76.5 | 22.6 | 0.9 |

Probably zinc was only isolated accidentally and occasionally in small amounts in antiquity in the course of smelting certain ores under special conditions, in much the same way as this metal was at first isolated accidentally in early modern times in Germany. In this connection it seems significant that the primary ore worked for silver in the famous ancient Greek mining district of Laurion was a mixture of the sulfides of iron, lead, and zinc.[6] It also seems significant that a product called *lauriotis*, evidently a form of zinc oxide named after the Laurion mining district, was obtained from the smelting furnaces there.[7]

[6] Marinos, G. P., and Petrascheck, W. E., *Laurion* (Athens, 1956). pp. 232–233.

[7] Pliny, *Natural History*, Book XXXIV, secs. 100, 132.

The great scarcity of metallic zinc in classical antiquity is also indicated by the paucity of information about it in the works of the writers of that period. Only two passages, both by Greek authors, can be interpreted as referring to the isolation of zinc, and one of these is apparently a fairly close quotation of the other. The original passage occurred in the *Philippica* of Theopompus, a historian of the fourth century B.C., but its occurrence in this work is known to us only through quotations of it or reference to it in the works of much later writers. It is, for example, quoted by Stephen of Byzantium, a writer of the sixth century A.D. in one of the surviving fragments of his *Geographical Lexicon.*[8] This writer also indicates that an earlier quotation of it is given by Strabo in his *Geography,* which was written about the beginning of the Christian Era. Strabo does not state or even hint that he quoted from Theopompus, but a comparison of the two quotations clearly shows that he did. The quotation by Strabo [9] may be closer to the original, for he was much nearer to Theopompus in time of writing and probably used an earlier manuscript of the *Philippica* than was available to Stephen of Byzantium. This quotation may be closely translated as follows: *There is a stone near Andeira which yields iron when burnt. After being treated in a furnace with a certain earth it yields drops of false silver. This, added to copper, forms the so-called mixture, which some call oreichalkos.* This brief passage, obviously a very sketchy account of a series of steps in a metallurgical process, has been much discussed, often as though it were originally written by Strabo. Some commentators appear to assume that this was the

[8] For a text of this quotation see: Grenfell, B. P., and Hunt, A. S., *Hellenica Oxyrhynchia cum Theopompi et Cratippi Fragmentis* (Oxford, 1909), *Philippica,* Book XII, sec. 109.

[9] *Geography,* Book XIII, sec. 56.

process, or one of the processes, used for the production of orichalcum around the beginning of the Christian Era. What they have overlooked, however, is that Strabo merely quotes from Theopompus, that he probably had no first-hand knowledge of metallurgy, and that the process may long since have fallen into disuse. In spite of various grammatical and metallurgical difficulties in this passage, it is generally agreed that the metal called false silver was more or less pure zinc, and that the final product called *oreichalkos* was an alloy containing copper and zinc as principal components. It seems difficult to avoid the conclusion that some kind of brass was made by this process in one locality from a particular ore as early as the fourth century. B.C. Possibly the process was inefficient and not productive of much *oreichalkos.* The scarcity of this alloy at that time is evident from the remarks of various Greek writers,[10] who commonly refer to it along with gold and silver in a way that indicates that it was both rare and costly.

Another brief passage, probably written about the third century B.C., appears to indicate that a cementation process for making a copper-zinc alloy was also used in Greek times. This passage comprises all of sec. 62 of the pseudo-Aristotelian compilation *De Mirabilibus Auscultationibus* (On Marvelous Things Heard). A close translation is as follows: *They say that the bronze of the Mossynoeci is very shiny and light in color, though tin is not mixed with the copper, but a kind of earth which occurs there is smelted with it. But they say that the discoverer of the mixing process did not instruct anyone else, so that the bronze objects formerly produced there are superior, whereas those made subsequently are not.*

[10] e.g., Plato, *Critias,* secs, 114e, 116b,d, 119, c-d.

Though other interpretations are possible, it seems very probable that this account refers to the manufacture of brass or zinc bronze by cementation. What seems to support this view is that the only known Greek objects composed of brass or zinc bronze (Table II) were found at or near the sites of colonies or settlements on the Black Sea, near the shores of which, in Pontus, dwelt the people known as the Mossynoeci. Even if this account does not allude to a process for making brass or zinc bronze, it is still important, as is likewise the one in sec. 49 of the treatise *On Stones* of Theophrastus,[11] for explaining how the manufacture of brass by cementation came to be discovered. Once the practice of heating various mineral substances with copper came into use it was inevitable that sooner or later zinc ores would be heated with the metal in the presence of a reducing agent and brass would be produced. What remains unknown, and probably will always remain unknown, is exactly where and when this discovery took place. Largely on the basis of the account just discussed, Forbes[12] believes that the discovery took place in Pcntus, but the evidence seems too incomplete to establish this with certainty. Forbes is also of the opinion that the discovery took place in the first half of the first millennium B.C., but real evidence for such an early date appears to be lacking. The available literary evidence does not indicate a date any earlier than the fourth century B.C., and the available archeological evidence indicates an even later date. As for the Roman orichalcum used for coinage, the earliest known date for a coin of this alloy is 45 B.C.

[11] For text, translation, and commentary see Caley, E. R., and Richards, J. F. C., *Theophrastus On Stones* (Columbus, 1956), pp. 26, 55, 162–167.

[12] Forbes, R. J., *Metallurgy in Antiquity* (Leiden, 1950), pp. 279–280.

Its manufacture on a large and continuous scale did not begin until shortly after the foundation of the Roman empire.

Though no explicit account of the process used by the Romans for the manufacture of orichalcum has come down to us, the hints given by Pliny[13] and by Dioscorides[14] leave no doubt that a cementation process was used. According to Pliny, the alloy was formed when copper was treated with a substance called *cadmea*. This word and the corresponding Greek *kadmeia* were, in the first century A.D., names that denoted two distinct groups of related substances. The first group included certain zinc minerals, probably only calamine and smithsonite, and the ores composed chiefly of one mineral or the other. Some statements of Pliny would appear to indicate that one kind of copper ore was also included, but this would seem to be one of his not uncommon errors. The second group included artificial products obtained as sublimates in the flues and on the walls of smelting furnaces, but these were sometimes prepared by roasting one of the substances in the first group separately in a special furnace. These artificial products appear always to have been forms of zinc oxide that differed only in physical form and degree of purity. Some of them were given special names based on their superficial appearance. Natural *cadmea* or *kadmeia* appears to have been the kind used to treat copper in order to convert it into orichalcum, but there is a possibility that the artificial products were sometimes used.

In the cementation process for the manufacture of brass as carried out in medieval and early modern times, bars, plates, or irregular pieces of copper were buried in a mix-

[13] *Natural History,* Book XXXIV, sec. 4.
[14] *Materia Medica,* Book V, Chapters 84–85.

ture of zinc ore and charcoal in a crucible, and on heating the charge to a sufficiently high temperature the zinc ore was reduced by the charcoal; and the liberated metal in the form of vapor was then absorbed by the copper. The degree of penetration of the copper by the zinc and the total proportion of zinc absorbed was dependent on the surface area of the copper, the thickness of the copper, the temperature, and the length of time of treatment, but it was not possible by this treatment alone to produce a uniform alloy. To do this, the temperature was raised to the fusion point and the molten metal was thoroughly stirred. A remark of Dioscorides indicates that the Roman process was different. He states that in the final treatment of copper the founders threw on it large portions of finely ground *kadmeia* in order to improve the quality of the metal. Since such treatment with a zinc ore or mineral alone would have had no effect, some reducing agent such as charcoal must also have been added. Probably the actual procedure was to stir both charcoal and *kadmeia* into the molten copper, possibly in the form of an intimate mixture of the two. Even though a copper-zinc alloy was so produced, a considerable proportion of the zinc ore or mineral must have been lost by volatilization. An indication of this is the mention by Dioscorides of the very fine white smoke (i.e., zinc oxide) that was evolved during the treatment of the copper. The quantity of this was so great that it was worth while to collect it from the flues and walls of the melting furnace. Not only was this process for the manufacture of orichalcum wasteful of the zinc ore or mineral, but the proportion of zinc in the finished alloy must have been difficult to control. However, neither the metal workers nor anyone else at that time had the idea that zinc was alloyed with copper in the process, and that the color and other properties varied with the proportion of zinc. The statements of both Pliny and Dioscorides show that they had no understanding of the chemical changes involved in the treatment of copper with compounds containing zinc. Pliny apparently believed that the metal absorbed the *cadmea*, and Dioscorides believed that the purpose of the treatment was merely to improve the quality of the copper.

A rather close estimate of the purity of the copper used in the manufacture of orichalcum may be obtained from the data of the new analyses of this investigation, for the impurities found on analysis are largely, if not entirely associated with the copper, as is indicated by analyses of Roman bronze and copper. On the assumption that all the impurities were associated with the copper and none with the zinc, the purity of the copper used in the cementation process may be calculated simply by prorating the percentages of copper and total impurities found on analysis over a scale of 100 percent. The average percentages of copper thus calculated from the analyses of early orichalcum coins are listed in Table IV. Since the As was struck from ordinary unalloyed copper, in all probability of the same quality as that used for making orichalcum, the proportion of copper in coins of this denomination should be a means of checking the correctness of this assumption. It will be seen from Table IV that the agreement is close. This means that the raw material containing zinc used in the manufacture of orichalcum was of rather high purity. In other words it was not crude zinc ore containing a marked proportion of metal impurities, but was at least carefully selected ore, and it may well have been a pure zinc mineral or even pure zinc oxide manufactured by sublimation.

TABLE IV.

Calculated Purity of the Copper Used in Making Early Coinage Orichalcum Compared With the Purity of Contemporaneous Unalloyed Roman Coinage Copper

| *Emperor* | *Number of Orichalcum Coins Analyzed* | *Calculated Purity of Copper Used in Manufacture of Orichalcum* % | *Copper Content of Contemporaneous Roman Copper Coins* % |
|---|---|---|---|
| *Augustus* | 2 | 98.75 | 98.73 |
| *Tiberius* | 1 | 99.53 | 99.65 |
| *Caligula* | 4 | 99.38 | 99.17 |
| *Claudius* | 3 | 98.45 | no data |
| *Nero* | 3 | 99.38 | 99.10 |
| | | Av. = 99.10 | Av. = 99.16 |

Various causes may account for the progressive decrease in the average zinc content of orichalcum coins beginning with the reign of Nero. One likely cause is that zinc minerals and ores were becoming increasingly scarce and costly because of the exhaustion of deposits, so that less came to be used in the manufacture of the alloy. An even more likely cause is the remelting of earlier worn coins in order to obtain metal for new coins. If worn *sestertii, dupondii,* and *asses* were melted together for this purpose the resulting alloy would, of course, have a lower zinc content than the worn coins. Even if only worn orichalcum coins, the *sestertii* and *dupondii,* were remelted, the resulting alloy would have a lower zinc content than the worn coins because of the preferential loss of zinc by volatilization and oxidation. Moreover, the occurrence of increasing proportions of tin and lead in orichalcum coins after the reign of Nero indicates that scrap bronze in some form was being added in the manufacturing process, probably in the remelting of coins. Possibly this bronze was chiefly in the form of worn Roman colonial coins. This addition of bronze increased the tin and lead content of the metal to such an extent that by the time of Commodus a majority of the *sestertii* and *dupondii* were being struck in zinc bronze instead of orichalcum. After his reign orichalcum practically disappeared as a coinage metal. In general the analytical data indicate that up to the time of Nero, orichalcum was always newly manufactured for coinage by a single standardized process, but that after his reign a large proportion, and perhaps at times all, of the alloy used for coinage was secondary metal. The manufacture of new orichalcum for coinage certainly ceased shortly after the reign of Antoninus Pius, and it may have ceased before that.

Since these conclusions are based entirely on analyses of orichalcum coins it might be supposed that they would not apply to the manufacture of Roman orichalcum in general. However, the archeological and numismatic evidence indicates that orichalcum was initially used solely for coins, and that its manufacture was a state monopoly.[15] There is a considerable likelihood that its manufacture continued to be a state monopoly, and that it was always manufactured exclusively for coinage. The relatively few orichalcum objects other than coins that have been found and analyzed have a composition very similar to that of the coins. Nearly all of these objects are small, their weight rarely exceeding that of a *sestertius*. It seems reasonable to conclude that orichalcum coins served as the immediate source of the metal for all these objects.

[15] Grant, M., *From Imperium to Auctoritas* (Cambridge, 1946), p. 88.

# Characteristics of Antiques and Art Objects by X-Ray Fluorescent Spectrometry

By Yoshimichi Emoto
Tokyo National Institute of Cultural Properties

The examination of materials of the cultural properties of fine and decorative art objects such as paintings and sculptures is necessary in making decisions relating to custody, preservation and repair of these cultural objects. It is also very useful in the study of cultural history, the history of the fine arts, archeology, and other disciplines that would benefit from improvements in determining dates of origin and techniques of production of art objects.

In Japan, however, it is rarely permitted to scrape sample materials for such analysis, and even if it is permitted, it is limited to very small amounts of particular parts of the objects at the time of repair. Otherwise, the examination must, for the most part, depend on broken or scaled-off fragments. Accordingly, inaccuracy in analysis using such fragments as samples, since they usually differ from the originals because of contamination or are not representative of the whole, is unavoidable. Thus, it has been necessary to devise non-destructive methods of qualitative examination suitable for these valuable objects. Hitherto the optical method, utilizing ultra-violet rays or infrared-rays, radioactivation analysis, the methods by measurement of $\beta$-ray back-scattering, X-ray radiography, and others have been practiced. X-ray-fluorescence analysis is considered to be the most effective method of measurement among these various non-destructive methods.

The writer has, in the last few years, been conducting qualitative examination of cultural objects by X-ray fluorescence analysis which has been remarkably developed in recent years, because it makes non-destructive and prompt analysis possible. If necessary, it is used together with the X-ray diffraction method.

Since previous determination methods have been more or less destructive, there are now unanalyzed materials. Some doubt whether what has been stated about these materials for a long time is really true or not. When they are subjected to analysis, they are sometimes seen to be of quite another character than what might have been expected.

It may not be too difficult to analyze these materials qualitatively. In speaking of quantitative analysis, on the other hand, there are many factors to be considered—the surface condition of the materials, the preparation of chemically standardized samples to be compared with the unknown materials, selection of historically standardized samples for each period. Above all, as we have stated, the objects to be analyzed must be guarded against damage, even at the expense of accuracy in the analysis. Thus, one is frequently obliged to have recourse to semi-quantitative methods of analysis.

Now, art objects are composed of a wide variety of materials. One encounters bronze, copper, gold, silver,

**Fig. 1. The five-story pagoda of Daigo Temtele in Kyoto. The pole at the top is the Sorin ornament.**

lead, iron and other metals; ceramics, glasses, pigments, etc. However, it has been proved that effective and non-destructive study is possible on all of these materials.

As to bronze, examining various parts of Sorin, an ornamental pole on the top of the five-story pagoda of the Daigo Temple in Kyoto (see Fig. 1), it was possible to make an approximate determination of the difference between the original parts of the ornament and the parts restored in later periods. This is done by the comparison of the impurity of Au, As and Sb. Quantitative studies on gold—sheet gold covering the ornament and the gold contained in Koban (gold coins)—were also made.[1] As to ceramics, a vase believed to have been produced in the Kamakura Era was shown to be a recently produced fake. (See Fig. 2). That is to say, a large difference in the intensity ratio, $SrK\alpha/RbK\alpha$ in the X-ray fluorescence spectrum of the glaze was shown between the product of the Kamakura Era and a contemporary one. This difference led to a conclusive distinction between the two after the comparative examination of many vases.

In the next stage the following research was carried out in order to find the degree of error in non-destructive analysis. Generally speaking, the factors that influence the accuracy of measurement are: surface condition of the samples, statistical fluctuation of X-ray, X-ray inter-element effect with the sample and the standardized materials. However, the unevenness of surface was taken into consideration.

To begin with, unsymmetrical reliefs were hammered out on the surface of the Ag-Cu alloy plate containing 10 percent (wt.) silver and 90 percent (wt.) copper in order

[1] Reported at the 14th Annual Congress of the Japan Chemical Society.

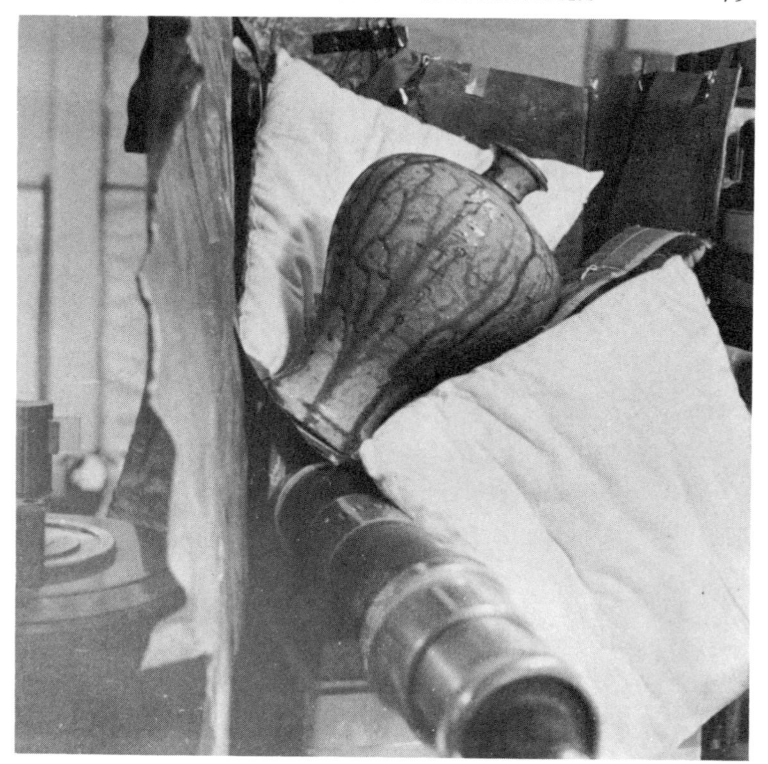

**Fig. 2. The forging vase prepared for examination by X-ray fluorescent spectrometry.**

to measure the unevenness of the surface. (See Fig. 3.) The writer reasoned that if measurement is made while rotating the sample, the unevenness would be averaged and the degree of accuracy would be increased. Therefore, comparative study was made between the measured value on the rotating sample and the averaged measured value on a sample which was not rotated but with the direction changed 45 degrees per measurement and turned through

360 degrees. In this case, the fluctuation ratio of each value of the latter was given against the value measured by flat plate of the same sample.

The result is, as shown in Fig. 4, that the value measured by the rotating sample corresponded well with the average value for each direction in the non-rotated sample. The

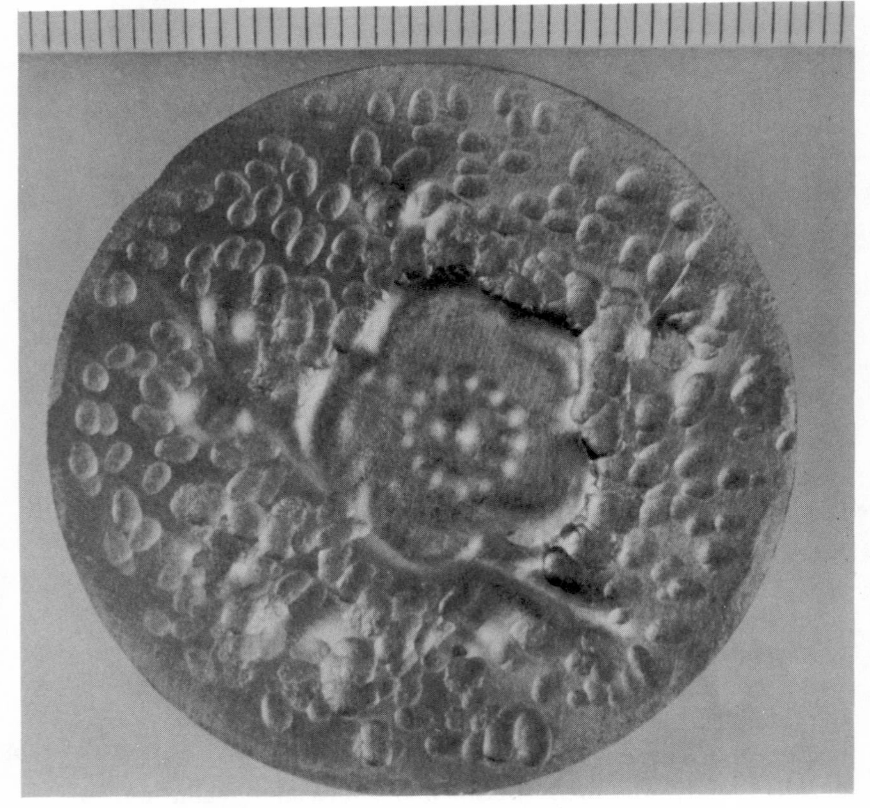

**Fig. 3. The sample plate on which unsymmetrical reliefs of plum blossom are hammered. Top inset shows the scale in millimeters.**

standard deviation for the rotated sample was 0.4 percent or less than one-fourth of the average value of the non-rotated sample for each direction.

The difference in deviation values between smooth and uneven samples converted into weight percentage, however, is approximately 5 percent; this difference is taken into account in determination of the chemical analytical value due to the surface unevenness of this particular sample.

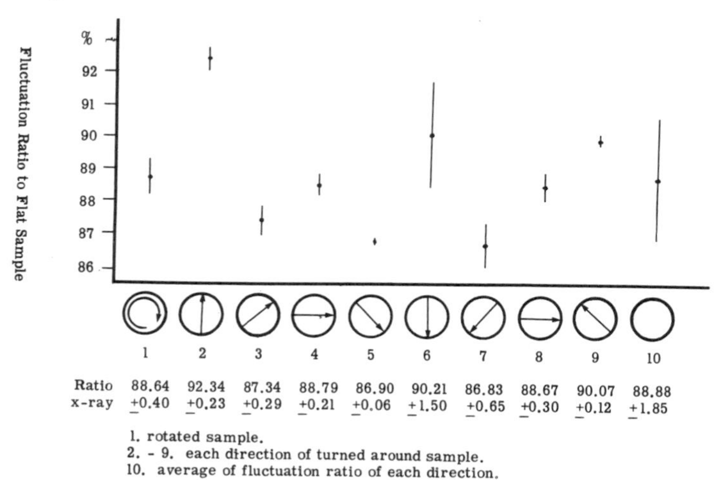


**Fig. 4. Effect of rotating sample.**

In conclusion, the condition of the surface is a difficult problem because of errors caused by unevenness, rust and other sources of irregularity. Examining how the conditions of such surfaces affect the accuracy of the measurement, the writer made an experiment on the problem of unevenness of the surface as a first attempt to decrease the

**Fig. 5. A decorative part of an old Japanese wooden building on the large sample holder for pigment identification.**

systematic error even a little and this time an approximately 5 percent (weight percentage) value was obtained.

This value is not inconsiderable and might be much larger if the effect of rust is added. However, it is possible to obtain this value and analytical content should be considered taking this systematic calculation of error into account. This is true with other non-destructive methods.

Systematic error cannot be avoided since the samples are very precious cultural objects which do not permit even of polishing the surface or making it even, but the writer believes the analysis is still worthwhile in gaining insights

**Fig. 6. Sassanian silver coins: 1. Hormazd IV. 2. Bahram VI. 3-5. Khusrau II.**

into ancient techniques or other historical problems. The writer is planning to make a similar examination of the materials in which unevenness is otherwise brought about or on the problem of rust.

By the method mentioned above of rotating the sample, the writer determined the content of copper and silver in Sassanian silver coins collected in Iran on the basis of calibration curves made on general standard samples. (See Fig. 6.)

*Apparatus and Conditions of Operation.* The Rigaku Denki X-ray fluorescence spectrometer with A Lif crystal, the Scintillation counter, and a Platinum target X ray tube (Machrett). The X-ray tube was operated at 25 kilovolts and 30 milliamperes for copper ($CuK\alpha$) and at 40 kilovolts and 10 milliamperes for silver ($AgK\alpha$). Fixed counts are 32,000 for copper and 256,000 for silver. The result is as shown in the Table.

The observed total content does not reach 100 percent because of the error caused by irregularities in the samples. Although it is rather difficult to find the date of origin or a local characteristic, due to the shortage of sample and a narrow range of its time period, the writer believes that these coins were made from silver produced at that time, and that copper was not intentionally mixed but was introduced as an impurity.

Since the institute purchased the apparatus only recently, the writer has not had a chance to collect enough data, and no significant findings have yet been obtained. However, the writer is making an effort to accumulate basic data useful for the history of fine arts, the history of technique and for archeological studies, analyzing samples which will become standard materials in various areas or samples for which the date of origin is clearly known.

Analytical Results on Sassanian Silver Coins

| *No.* | *Kings' Name and Chronology* | | *Ag* | | *Cu* | | *Total* |
|---|---|---|---|---|---|---|---|
| | | | *I/I°* | *Ag %* | *I/I°* | *Cu %* | |
| 1 | Kobad I | 488-531 A.D. | 88.99 | 90.5 | 12.85 | 3.2 | 93.7 |
| 2 | Khusrau I | 30th yr. 561 A.D. | 87.89 | 89.5 | 112.16 | 3.0 | 92.5 |
| 3 | Hormazd IV | 9th yr. 587 A.D. | 81.76 | 84.0 | 6.04 | 1.5 | 85.5 |
| 4 | 〃 | 579-590 A.D. | 92.47 | 93.5 | 4.94 | 1.2 | 94.7 |
| 5 | 〃 | 〃 | 87.04 | 88.8 | 7.62 | 1.9 | 90.7 |
| 6 | Bahram VI | 590-591 A.D. | 89.34 | 91.0 | 8.82 | 2.2 | 93.2 |
| 7 | 〃 | 〃 | 90.47 | 91.7 | 3.13 | 0.8 | 92.5 |
| 8 | Khusrau II | 5th yr. 594 A.D. | 89.03 | 90.5 | 7.11 | 1.8 | 92.3 |
| 9 | 〃 | 8th yr. 597 A.D. | 87.35 | 89.1 | 4.61 | 1.2 | 90.3 |
| 10 | 〃 | 11th yr. 600 A.D. | 88.20 | 89.8 | 7.00 | 1.8 | 91.6 |
| 11 | 〃 | 〃 | 89.05 | 90.5 | 4.91 | 1.2 | 91.7 |
| 12 | 〃 | 〃 | 84.48 | 86.5 | 6.56 | 1.7 | 88.2 |
| 13 | 〃 | 25th yr. 614 A.D. | 85.70 | 87.7 | 12.21 | 3.2 | 90.9 |
| 14 | 〃 | 590-628 A.D. | 90.51 | 92.0 | 11.52 | 2.9 | 94.9 |
| 15 | 〃 | 〃 | 88.28 | 90.0 | 3.89 | 1.0 | 91.0 |
| 16 | 〃 | 〃 | 85.27 | 87.3 | 9.67 | 2.5 | 89.8 |
| 17 | 〃 | 〃 | 90.74 | 93.0 | 8.30 | 2.1 | 95.1 |
| 18 | 〃 | 〃 | 87.84 | 89.5 | 11.40 | 2.9 | 92.4 |
| 19 | 〃 | 〃 | 90.74 | 92.2 | 8.91 | 2.2 | 94.4 |
| 20 | 〃 | 〃 | 82.84 | 85.1 | 8.91 | 2.2 | 87.3 |
| 21 | 〃 | 〃 | 84.77 | 86.7 | 13.07 | 3.3 | 90.0 |
| 22 | Unknown | 687 A.D. | 80.13 | 82.5 | 14.85 | 3.7 | 86.2 |
| 23 | 〃 | Unknown | 84.99 | 87.0 | 3.04 | 0.8 | 87.8 |
| 24 | 〃 | 〃 | 89.68 | 91.2 | 14.85 | 3.7 | 94.9 |
| 25 | 〃 | 〃 | 88.50 | 90.2 | 8.94 | 2.2 | 92.4 |
| 26 | 〃 | 〃 | 85.40 | 87.5 | 3.24 | 0.8 | 88.3 |
| 27 | 〃 | 〃 | 90.74 | 92.2 | 8.91 | 2.2 | 94.4 |
| 28 | 〃 | 〃 | 86.28 | 88.2 | 5.41 | 1.4 | 89.6 |

# Chemische Untersuchungen der Patina vorgeschichtlicher Bronzen aus Niedersachsen und Auswertung ihrer Ergebnisse

Von Wilhelm Geilmann, Mainz

Die Übertragung der in den letzten Jahrzehnten entwickelten Verfahren der Materialprüfung auf die Untersuchung der kulturgeschichtlich interessanten Fund-und Sammlungsstücke kann Erkenntnisse vermitteln, die wesentlich über die hinausgehen, die bei alleiniger Berücksichtigung des äußeren Aussehens möglich sind und kann Einblicke in das technische Können and Wissen vergangener Zeitepochen ergeben, die auf keinem anderen Wege zu erlangen sind.

Auch für die in den meisten Fällen erforderlichen Konservierungs-verfahren kann eine solche Untersuchung recht bedeutungsvoll sein, da sie ein recht zuverlässiges Bild vom Erhaltungszustand des Stückes selbst und der Gegenwart oder dem Fehlen der die spätere Erhaltung beeinträchtigenden Stoffe liefert, können die am besten geeigneten Arbeitsverfahren viel besser und sicherer ausgewählt werden, als es sonst möglich sein würde.

Die vor allem von Kunstgeschichtlern häufig geäußerte Befürchtung, daß bei einer derartigen Untersuchung mit einer Schädigung der Stücke zu rechnen sein würde, ist heute keinesfalls mehr stichhaltig, da eine Reihe von Arbeitsverfahren ohne jede Materialzerstörung durchführbar sind.

Auch die moderne chemisch- analytische Untersuchung ist gegenüber den älteren Methoden soweit verfeinert, daß sie mit Substanzmengen von wenigen Milligrammen durchführbar ist und dabei zu Ergebnissen führt, die in ihrer Zuverlässigkeit die übertreffen, die in früheren Zeiten Probemengen im Gewichte von Grammen verlangten. Die heute für Mikroanalysen benötigten Milligramme und oft noch weniger, sind aber ohne merkliche, vor allem ins Auge fallende Beschädigung zu antnehmen, sodaß von einer Beeinträchtigung auch kostbarer Fundstücke kaum die Rede sein kann.

Im übrigen sei darauf hingewiesen, daß die bei der Freilegung oder Präparierung der Fundstücke abfallende Teile, die meistens unbeobachtet bleiben oder gar verworfen werden, in vielen Fällen Material darstellen, dessen Untersuchung zu recht wertvollen Erkenntnissen führen kann.

## UNTERSUCHUNGEN AN METALLEN

### a. Patinauntersuchungen.

Bei dem Lagern im Boden überziehen sich Metalle und Legierungen durch die Einwirkung von Luft, Wasser, Kohlendioxyd und den in den Bodenlösungen vorhandenen Salzen mit einer Patinaschicht, deren Zusammensetzung abhängig ist vom Metall und der Zusammensetzung der mit ihm in Berührung gekommenen Lösungen.

Auf den im wesentlichen interessierenden antiken Bronzen, also Kupferzinnlegierungen, können die nachstehend aufgeführten Verbindungen vorkommen, die allein, im Gemisch miteinander oder scharf von einander getrennt, auftreten.

1. Das häufigste Patinamaterial ist der grüne Malachit, ein basisches Kupferkarbonat der Formel ($CuCO_3.Cu(OH)_2$). Wesentlich seltener und dann nur in feinsten Nadeln im Malachit eingebettet tritt ein anderes basisches Karbonat auf, die blaue Kupferlasur ($CuCO_3.Cu(OH)_2$).
2. An den Übergangsstellen vom Metall zur grünen Karbonatpatina sind die Oxyde des Kupfers vorhanden und im Anschliff mikroskopisch gut zu erkennen, der rote Cuprit ($Cu_2O$) und der schwarze Tenorit ($CuO$).
3. Seltener tritt ein basisches Chlorid des Kupfers, der Atakamit $CuCl_2. 3Cu(OH)_2$ auf. Er findet sich in stark schwankender Menge in jeder Patinaschicht, die unter Einwirkung stärker chloridhaltiger Lösungen entstanden ist. Reichlich findet er sich in der Patina aegyptischer Bronzen, die der Einwirkung der aus den Wüstengebieten stammenden Salzlösungen ausgesetzt war. In geringer Menge erscheinen Chloride auch in der Patina der Bronzen anderer Herkunft, ihre Anwesenheit ist besonders störend, da sie Veranlassung zur fortschreitenden Korrosion beim Aufbewahren an der Luft mit ihrem schwankenden Feuchtigkeitsgehalt geben.

Die Prüfung der Patina auf Chloride und deren vollständige Entfernung ist aus diesem Grunde für die Erhaltung eines Fundstückes von grundsätzlicher Bedeutung.

4. Basisches Kupfersulfat von smaragdgrüner bis schwärzlichgrüner Farbe, als häufiges Mineral in der Oxydationszone von Kupfererzen in Trockengebieten unter dem Namen Brochantit bekannt, tritt in Bronzefunden aus humiden Gebieten nur spurenweise auf, dagegen kann es

auf neuzeitlichen, den Rauchgasen mit höherem $SO_2$-Gehalt ausgesetzten Bronzen, Kupferdächern und Drähten in reichlicherem Maße zu erwarten sein.
5. Basisches Kupferarsenat, $Cu_3(AsO_4)_2.Cu(OH)_2$ von laucholivgrüner Farbe ist als Olivenit bekannt und dürfte nur in der Patina stärker Arsen-haltiger Bronzen zu erwarten sein.
6. Ein garnicht seltenes, aber kaumerwähntes lauch-bis olivgrün gefärbtes Verwitterungsprodukt auf Bronzen ist der Libethenit, ein basisches Phosphat der Zusammensetzung $Cu_3(PO_4)_2.Cu(OH)_2$, das entsteht sobald Kupferlösungen mit Knochen in Berührung kommen. 1)
7. Das wesentliche Verwitterungsprodukt des Zinns ist die gelförmige Zinnsäure, $SnO_2.\ xH_2O$, die auch mikroskopisch schwer zu erkennen ist und sich in der Patina infolge ihrer Schwerlöslichkeit erheblich anreichern 2) und durch ihre starke Adsorptionsfähigkeit aus den Bodenlösungen eine Reihe von Stoffen aufnehmen kann.

*1. Untersuchung einiger Patinaschichten.* Für die Analyse wurde nur die Patina verwandt, die durch Abklopfen und Abkratzen von den Bronzen aus niedersächsischen Fundgebieten erhalten war, wobei trotz aller Vorsicht eine gelegentliche geringe Verunreinigung mit dem dem noch unzersetzten Metall unmittelbar aufliegenden, fest haftendem Kupferoxydul unvermeidbar war. Hier wird der angegebene CuO Gehalt um höchstens einige 1/10% zu hoch erscheinen, was aber für die Auswertung der Analysen belanglos sein dürfte.

Die Ergebnisse der Untersuchung sind in Tabelle 1 zusammengestellt Bei der verwandten Analysentechnik dürften die angegebenen Befunde mit Fehlern von höchstens $\pm$ 0, 05% behaftet sein, umsomehr, da sie meistens Mittelwerte aus mehreren Bestimmungen sind.

TABELLE 1.

Patina-Analysen

| *Nr.* | 1 | 2 | 3 | 4 | 5 | 6 | 7 | 8 | 9 | 10 | 11 |
|---|---|---|---|---|---|---|---|---|---|---|---|
| | % | % | % | % | % | % | % | % | % | % | % |
| CuO | 24.89 | 39.94 | 50.84 | 57.20 | 63.09 | 45.76 | 69.26 | 73.07 | 67.85 | 89.23 | 75.78 |
| $SnO_2$ | 57.12 | 24.09 | 14.75 | 6.78 | 6.20 | 3.65 | 4.30 | 0.78 | 0.65 | 0.92 | 0.42 |
| $Sb_2O_5$ | — | — | — | 0.10 | 0.15 | 0.05 | — | — | 0.05 | — | — |
| PbO | — | — | Spur | — | 0.37 | — | Spur | 0.05 | — | — | 0.15 |
| $Fe_2O_3$ | 1.71 | 1.03 | 0.75 | 3.86 | 0.95 | 1.55 | 0.10 | 0.78 | 0.29 | 0.96 | 0.25 |
| CaO | — | 1.75 | 0.11 | — | 0.86 | 0.34 | 0.13 | — | — | — | — |
| $P_2O_5$ | — | 0.15 | 2.93 | Spur | 1.25 | 1.92 | 1.77 | 0.24 | 0.01 | 0.63 | 0.18 |
| $SO_3$ | — | — | Spur | — | — | Spur | — | — | — | — | 0.05 |
| Cl | Spur | Spur | Spur | 0.05 | 0.15 | 0.08 | Spur | 0.02 | — | 0.28 | Spur |
| Ni | Spur | — | — | — | — | Spur | — | 0.01 | 0.05 | — | 0.05 |
| Glühverlust | 14.35 | 32.52 | 29.55 | 32.20 | 27.14 | 46.72 | 24.44 | 26.08 | 31.07 | 7.39 | 23.07 |
| Summe | 98.07 | 99.48 | 98.93 | 100.19 | 100.16 | 100.07 | 100.00 | 101.03 | 99.97 | 99.41 | 99.95 |
| Verhältnis | 0.44 | 1.68 | 3.50 | 8.28 | 10.08 | 12.72 | 16.24 | 94.74 | 98.31 | 98.40 | 183.7 |
| Cu/Sn | 1 | 1 | 1 | 1 | 1 | 1 | 1 | 1 | 1 | 1 | 1 |

Für die Untersuchung wurden die nachfolgend aufgeführten Bronzen verwandt, die aus Beständen des Niedersächsischen Landesmuseums zur Verfügung gestellt wurden.

1. Ein Absatzbeil, aus Ricklingen bei Hannover. 2. Ein Beil aus Leveste. 3. Dolchklinge, Westerwanna. 4. Lanzenspitze, Neetze. 5. Schwertklinge aus Bostelwiebeck. 6. Schwertklinge, Bergen. 7. Schwert griff, Bramstedt. 8. Lanzenspitze aus Westergellersen bei Lüneburg. 9. Lanzenspitze, Oldendorf. 10. Lanzenspitze Beverbeck. 11. Armreif, unbekannt.

In der letzten Zeile der Tabelle ist das aus den Oxyden berechnete Verhältnis von Kupfer zu Zinn aufgeführt, wobei der Wert für Zinn gleich 1 gesetzt wurde.

Eine derartige Vergleichsberechnung dürfte ohne weiteres zulässig sein, da auf Grund des chemischen Verhaltens der entstandenen Zinnsäure eine merkliche Auflösung und Fortführung durch die Bodenlösungen nicht zu erwarten ist. Bei den verwandten Bronzen schwankt dieses Verhältnis zwischen 8 zu 1 bis 11 zu 1. In der Patina wird es aber nur bei den Proben 4–6 erreicht, während es bei 1–3 wesentlich tiefer und bei 7–11 höher liegt, was bedeuten würde, daß in den Proben 1–3 ein Verlust an Kupfer und den Proben 7–11 eine Zunahme an Kupfer festzustellen ist.

*2. Zusammensetzung von Metall und Patinaschicht.* Wesentlich deutlicher als aus den alleinigen Patinaanalysen geht die beim Übergang vom Metall in die Patina eintretende Verschiebung im Metallgehalt aus dem Vergleich der Analysenzahlen für Metall und zugehöriger Patina hervor, wie die Zusammenstellung der zusammengehörigen Werte der Tabelle 2 lehrt.

Untersucht wurde das Metall und die Patina folgender Fundstücke.

Probe. 1. Ein Bronzebeil von nordischem Typ aus der Gegend von Soltau Niedersachsen, zeitlich einzuordnen 1500–1100 v. Chr. Eine leuchtend grüne, lockere Patina umgibt in 5–7mm starker Schicht den unregelmäßig und stellenweise lochartig angefressenen Metallkern, der, wie die mikroskopische Betrachtung zeigt, Gußstruktur aufweist und am Übergang zur Patina reichlich $Cu_2O$ erkennen läßt, das in das Metall zackig einwächst.

Probe 2. Ein 5.6–6.5mm starker Bronzereif von eliptischem Querschnitt, unbekannter Herkunft, jedoch aus Niedersachsen stammend. Der innere Metallkern von 3–4mm Stärke ist mit einer etwa ½mm starken roten $Cu_2O$ Schicht überzogen, von dem die öberflächlich glatte, stark glänzende, hellgrüne Patina in schalenförmigen Krusten abspringt. Untersucht wurde die von Resten $Cu_2O$ befreite Patina und das ebenfalls von ihm befreite Metall, das noch weich und hämmerbar war.

Probe 3. Bronzemesser aus der Zeit um 1000 v. Chr., gefunden bei Urbach, Kreis Ilfeld.

Das recht gut erhaltene Metall ist von einer festhaftenden roten Schicht von $Cu_2O$ umgeben, auf der eine fest haftende, dunkelgrüne und etwa ½mm starke Patinaschicht aufliegt, die stellenweise warzenförmige Ausblühungen aufweist.

Probe 4. Ein Armreif aus einem Körpergrabe bei Vorwohlde.

Der spiralförmige Armreif von dreieckigem Profil umschließt noch den Unterarm der Bestatteten und ist mit einer ½mm starken, fest haftenden, bräunlichgrünen Patina bedeckt. Der Metallkern ist brüchig und mit $Cu_2O$ durchsetzt.

TABELLE 2.

Analysen von Metall und Patina

| | *Metall* | | | | | *Patina* | | | |
|---|---|---|---|---|---|---|---|---|---|
| | 1 | 2 | 3 | 4 | | 1 | 2 | 3 | 4 |
| Cu | 90.80% | 87.12% | 89.58% | 85.40% | CuO | 30.35% | 23.05% | 38.85% | 48.70% |
| Sn | 7.63 | 11.82 | 9.52 | 10.92 | $SnO_2$ | 44.14 | 55.41 | 22.25 | 18.61 |
| Sb | 0.45 | 0.23 | 0.22 | 0.21 | $Sb_2O_5$ | 2.87 | 1.08 | 0.48 | 0.42 |
| As | 0.27 | 0.09 | 0.12 | n.b. | $As_2O_5$ | 1.47 | 0.48 | 0.31 | n.b. |
| Pb | 0.47 | 0.01 | 0.05 | Spur | PbO | 0.17 | — | Spur | — |
| Fe | 0.17 | 0.26 | 0.38 | 0.87 | $Fe_2O_3$ | 1.41 | 0.62 | 0.94 | 1.85 |
| Ni | 0.16 | 0.48 | 0.14 | Spur | NiO | — | 0.03 | Spur | — |
| Co | 0.01 | 0.04 | 0.02 | — | CoO | — | — | — | — |
| | 99.96 | 100.05 | 100.03 | 97.40 | $Al_2O_3$ | 0.25 | 0.21 | 0.34 | 0.59 |
| | | | Spur Ag, Zn, Bi | viel Sauerstoff u. etwas $CO_2$ | CaO | 0.05 | 0.05 | 0.07 | — |
| | | | | | $SiO_2$ | 0.17 | 0.22 | 14.94 | 6.53 |
| | | | | | $P_2O_5$ | Spur | 3.67 | 3.57 | 1.36 |
| | | | | | Cl | 0.18 | 0.10 | 1.11 | — |
| | | | | | $SO_3$ | 0.94 | 0.14 | 0.25 | 0.15 |
| | | | | | $CO_2$ | 6.95 | 5.83 | 10.53 | 22.08* |
| | | | | | $H_2O$ | 11.90 | 9.23 | 6.40 | — |
| | | | | | Humus | 0.02 | Spur | fehlt | 0.08 |
| | | | | | Summe | 100.77 | 100.13 | 100.04 | 100.37 |
| Cu/Sn | 11.9<br>1 | 7.37<br>1 | 9.41<br>1 | 7.82<br>1 | Cu/Sn | 0.70<br>1 | 0.43<br>1 | 1.77<br>1 | 2.65<br>1 |
| Sn/Sb | 16.9<br>1 | 52.1<br>1 | 43.3<br>1 | 32.0<br>1 | Sn/Sb | 16.1<br>1 | 53.7<br>1 | 48.1<br>1 | 46.3<br>1 |
| Sn/As | 28.3<br>1 | 13.1<br>1 | 79.2<br>1 | — | Sn/As | 33.8<br>1 | 14.6<br>1 | 90.7<br>1 | — |
| Sn/Fe | 44.9<br>1 | 45.4<br>1 | 25.0<br>1 | 12.5<br>1 | Sn/Fe | 35.2<br>1 | 10.0<br>1 | 26.7<br>1 | 11.3<br>1 |

Neben den Analysenwerten sind in den letzten 4 Reihen die errechneten Verhältniszahlen von Cu/Sn, von Sn/Sb, von Sn/As und von Sn/Fe aufgeführt.

Aus dem Vergleich der Analysenzahlen für das Metall und denen der zugehörigen Analysenzahlen ergibt sich:
1. Der Gehalt von Kupfer ist geringer geworden, was sich besonders deutlich aus den auf die Zinnmenge bezogenen Verhältniszahlen erkennen läßt.
2. Der Gehalt an Antimon und Arsen hat prozentual in der Patina scheinbar zugenommen. Die Verhältniszahlen zum Zinn sind jedoch beim Antimon nahezu konstant geblieben, die Schwankungen liegen innerhalb der durch die Bestimmungfehler bedingten Grenzen. Beim Arsen sind die Abweichungen scheinbar größer, aber im wesentlichen auch verursacht durch die unvermeidlichen Bestimmungsfehler, die sich bei den wesentlich geringeren Gehalten noch stärker als beim Antimon auswirken. Unter Berücksichtigung dieser Tatsachen kann auch der Arsengehalt als konstant gelten.

Beim Eisen ist bei den Proben 1 und 2 der Gehalt in der Patina höher als im Metall, was nur auf eine Zufuhr durch die eisenhaltigen Bodenlösungen zurückgeführt werden kann. Im Gegensatz hierzu ist der Eisengehalt der Patinaproben 3 und 4 allein durch den des Metalles bedingt.

Völlig verschwunden ist der geringe Gehalt an Blei, Nickel und Kobalt beim Übergang des Metalles in die Patina.

Die übrigen in der Patina auftretenden Stoffe, vor allem die Phosphorsäure, müssen aus der Bodenlösung zugeführt sein, das gilt auch für $SiO_2$, das als feinster Sand mit der Patina verkittet ist.

*3. Untersuchungen an Proben mit fortschreitender Zersetzung.* In den Sammlungen einzelner Museen finden sich, wenn auch nicht häufig, Bronzen, bei denen sich infolge günstiger Lagerung alle Zersetzungsstufen von reinem Metall bis zu der, dem weiteren Abbau widerstehenden Zinnsäure, dem Endprodukt der Verwitterung der Bronze, beobachten lassen.

Die chemische Untersuchung solcher Stücke mußte einen besonders guten Einblick in den Ablauf der Verwitterung ermöglichen.

Zur Verfügung standen zwei Proben, eine in einzelnen Bruchstücken vorliegende Schwertklinge und Reste eines Bronzegefäßes, das am Halsteil völlig zersetzt war, am Boden aber noch unangegriffenes Metall aufwies.

Das Schwert gehört der mittleren Bronze, also der Zeit um 1000 v. Chr. an und wurde in Wölpe geborgen. Es ist so vorzüglich erhalten, daß selbst alte Kratzer und Schlagmarken auf der Oberfläche erkennbar sind, selbst auf den Stellen, die restlos verwittert sind.

Im Griffteil findet sich im Innern völlig erhaltenes Metall, das von einer Oxydzone umschlossen ist, die in eine Patinaschicht übergeht, an die sich eine schwach gelbliche Schicht mit tief braunem Rand anschließt, wie aus der schematischen Zeichnung im Bilde 1 zu erkennen ist. Zur Schwertspitze hin folgt eine bis auf einen schwachen braunen Rand gleichmäßig grün gefärbte Zone, in der auch bei mikroskopischer Prüfung weder Metall noch Oxydreste erkennbar sind. An der Klingenspitze ist die Masse gelblich weiß gefärbt und die braune Randzone kräftiger ausgebildet.

Zur Untersuchung kamen Teile der Klingenspitze, Teile des mittleren Klingenteils und solche des Griffes. Nach sorgfältiger Trennung wurden von letzterem einzeln untersucht der Metallkern, die Oxydzone und gemeinsam, da sich eine scharfe Trennung als unmöglich erwies, die sich den Oxyden anschließende Patina-Zinnsäureschicht.

Die Analysenbefunde enthält die Tabelle 3 neben den bei der Untersuchung der nachstehend beschriebenen Bronzeurne erhaltenen Werte.

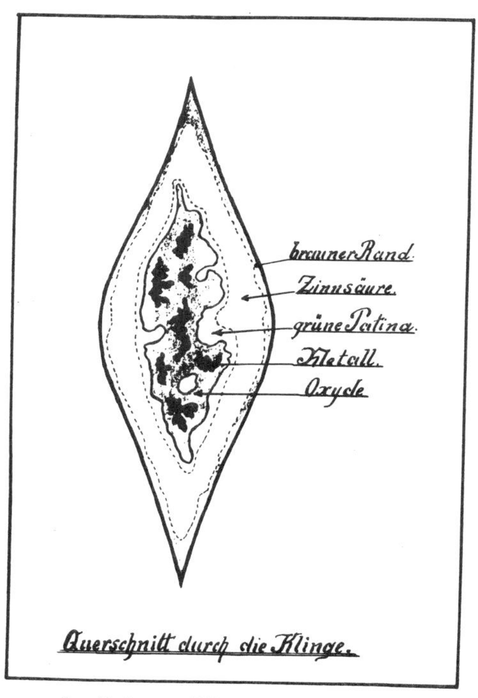


**1. Schwert Wolpe, Querschnitt.**

Das als Urne benutzte Bronzegfäß römischer Herkunft entstammt einem bei Helzendorf, Krs. Syke 3) gelegenem Grabe des 2.–3. Jahrh. n. Chr. Am Boden des Gefäßes ist ein 0.6–0.7mm starkes Metallblech unverändert erhalten und nur mit einer grünen, leicht absprigenden Patina bedeckt. Zum Halse des Gefäßes aufsteigend ist das ursprüngliche Blech in eine blaugrüne, sehr feste und glatte

Patinamasse umgewandelt, in der weder Oxyd noch Metallreste zu erkennen sind. Noch weiter zum oberen Rande hin verschwindet ziemlich schnell die Grünfärbung und dafür erscheint eine schwach grüne, feste Patinamasse, die Form und Dicke des ursprünglichen Metalls völlig beibehalten hat.

Die Verschiebung der Bestandteile geht bereits aus diesen Prozentzahlen hervor und steht völlig im Einklang mit den beim Übergang des Metalles zur Patina beobachteten Ergebnissen.

Wesentlich besser läßt sich die bei fortschreitender Verwitterung auftretende Verschiebung der eizelnen Bestandteile verfolgen, wenn sie auf Zinn als Bezugselement berechnet werden, dessen Menge konstant bleibt. Die dem Metall entstammenden Oxyde wurden zunächst auf die Elemente berechnet und dann durch den zugerhörigen Zinnwert geteilt und mit 100 multipliziert, um zu nicht zu kleinen Zahlen zu kommen. Bei den in die Patina eingewanderten Oxyden unterblieb die Umrechnung auf die Elemente, sie wurden unmittelbar auf 100Tle.Zinn bezogen.

Die so gefundenen Zahlen finden sich in der Tabelle 4, sie bedeuten, wieviel Teile Metall oder Oxyd jeweils auf 100g Zinn bezogen in der Probe vorhanden sind.

Bei einer derartigen Darstellung erkennt man sofort, daß Kupfer beim Übergang des Metalles zum Oxyd sich bereits wesentlich vermindert und in der Patina noch weniger erhalten bleibt. Mit fortschreitender Einwirkung der Bodenlösungen nimmt das Kupfer weiter bis zu einem Grenzwert ab, der dann konstant bleibt. Das Gleiche gilt für das Blei, Nickel, und Kobalt, infolge ihrer geringeren Konzentration verschwinden sie schneller bis auf einige 1/1000–1/100%, die spektrographisch in der "reinen

TABELLE 3.

Zusammensetzung einzelner Zonen bei Bronzen mit fortschreitender Zersetzung

| | *Schwertklinge* | | | | *Klingen* | |
|---|---|---|---|---|---|---|
| | *Metall* | | *Oxyd* | *Patina* | *Mitte* | *Spitze* |
| | % | | % | % | % | % |
| Cu | 90.26 | CuO | 72.02 | 22.55 | 8.72 | 5.60 |
| Sn | 8.70 | $SnO_2$ | 28.53 | 55.07 | 68.59 | 68.15 |
| Sb | 0.20 | $Sb_2O5$ | 0.75 | 1.34 | 1.46 | 1.74 |
| As | 0.05 | $As_2O_5$ | 0.42 | 1.09 | 0.93 | 0.91 |
| Pb | 0.18 | PbO | 0.01 | 0.05 | 0.05 | 0.05 |
| Fe | 0.15 | $Fe_2O_3$ | 0.12 | 0.72 | 0.91 | 1.64 |
| Ni | 0.24 | NiO | 0.07 | 0.15 | — | — |
| Co | 0.10 | CoO | 0.06 | 0.09 | — | — |
| | | $Al_2O_3$ | 0.01 | 0.15 | 0.25 | 0.52 |
| | 99.88 | $MnO_2$ | — | — | 0.05 | 0.12 |
| | | CaO | — | 0.05 | 0.10 | 0.09 |
| | | $P_2O_5$ | 0.12 | 1.01 | 1.07 | 1.05 |
| | | $SO_3$ | — | — | 0.12 | 0.05 |
| | | $SiO_2$ | 0.02 | 0.06 | 0.08 | 0.05 |
| | | $CO_2$ | 0.22 | 2.40 | 0.64 | 0.42 |
| | Glühverl | | — | 15.41 | 17.28 | 19.83 |
| | | | 102.35 | 100.14 | 100.20 | 100.19 |
| | Humus | | — | 0.16 | 0.15 | 0.49 |
| | $NH_4$ | | — | 0.03 | 0.06 | 0.08 |

| | *Bronzegefäb* | | | | |
|---|---|---|---|---|---|
| | *Metall* | | *Patina* | *Zinn-säure* | *Zinn-säure* |
| | % | | % | % | % |
| Cu | 83.70 | CuO | 54.84 | 9.30 | 0.78 |
| Sn | 11.85 | $SnO_2$ | 28.31 | 56.07 | 63.20 |
| Sb | 0.53 | $Sb_2O_5$ | 1.91 | 2.35 | 2.80 |
| As | 0.14 | $As_2O_5$ | 0.40 | 0.78 | 0.90 |
| Pb | 2.98 | PbO | 0.27 | 0.20 | 0.01 |
| Fe | 0.96 | $Fe_2O_3$ | 1.90 | 2.42 | 2.67 |
| | 100.16 | NiO | — | — | — |
| | | CoO | — | — | — |
| | | $Al_2O_3$ | 0.05 | 2.24 | 2.75 |
| | | $MnO_2$ | — | 0.09 | 0.15 |
| | | CaO | 0.05 | 0.10 | 0.03 |
| | | $P_2O_5$ | 0.76 | 2.39 | 2.69 |
| | | $SO_3$ | 0.10 | 0.05 | 0.02 |
| | | $SiO_2$ | 0.37 | 3.17 | 3.20 |
| | | $CO_2$ | 6.45 | 1.58 | 0.05 |
| | Glühverl | | 6.32 | 19.31 | 21.30 |
| | | | 99.83 | 100.05 | 100.25 |
| | Humus | | Spur | Spur | 0.08 |
| | $NH_4$ | | — | — | — |
| | | Cl | 0.05 | 0.08 | 0.03 |

Zinnsäure" meistens noch nachweisbar sind. Arsen und Antimon bleiben quantitativ im Verwitterungsrückstand.

Eisen bildet einen Sonderfall dadurch, daß im zunächst gebildeten Oxyd-Gemisch seine Menge sich deutlich vermindert, um später wieder anzusteigen. Dies Verhalten ist verständlich, wenn bedacht wird, daß bei der Bildung der Oxyde das Eisen zweiwertig vorliegen muß, das ähnlich den andern zweiwertigen Metallen fortgelöst wird. Erst bei Sauerstoffzutritt erfolgt die Oxydation und Abscheidung als Eisen (III) hydroxyd, das nicht mehr löslich ist und sich in der Patina abscheidet, wo sein Gehalt konstant bleibt.-durch Einwanderung noch erhöht werden kann.

Auch die Variation der von außen in die Patina einwandernden Stoffe ist besser zu verfolgen, als es bei den prozentischen Analysenzahlen der Fall ist.

Besonders hingewiesen sei auf das Verhalten der Phosphorsäure, die aus den verwitternden Knochen bezw. dem Leichenbrand löslich geworden in der Patina und der Zinnsäure in wesentlicher Menge festgelegt und im Rückstand verbleibt.

*4. Untersuchung der Verwitterungsreste von Bronzen.* Das aus den für die Zersetzung der Bronzen besonders günstige Böden, den nordwestdeutschen humosen Sandböden, schon frühzeitig völlig zersetzte Bronzen geborgen und in Sammlungen aufbewahrt wurden, dürfte selbstverständlich sein. In Verkennung der Tatsachen wurden sie als "Fundstücke aus Knocken oder Pfeifenton in der einer Bronze ähnlichen Form" bezeichnet-Bereits 1880 erkannte O. v. Olshausen 5) auf Grund seiner chemischen Prüfungen die wirklichen Zusammenhänge und bezeichnete diese Stücke, die die ursprüngliche Form vollständig beibehalten haben, als "wahre Pseudomorphosen von einer Zinnsäure

TABELLE 4.
Verhältniszahlen der Metalle bezw. Oxyde, bezogen auf 100 Teile Zinn

| | *Schwertklinge* | | | *Klingen* | | *Bronzegefärb* | | *Zinnsäure* | |
|---|---|---|---|---|---|---|---|---|---|
| | *Metall* | *Oxyde* | *Patina* | *Mitte* | *Spitze* | *Metall* | *Patina* | *a* | *b* |
| Cu | 1037.50 | 256.05 | 241.53 | 12.90 | 8.33 | 706.20 | 196.40 | 16.81 | 1.25 |
| Sb | 2.30 | 2.64 | 2.45 | 2.14 | 2.57 | 4.47 | 4.46 | 4.44 | 4.24 |
| As | 0.58 | 1.22 | 1.64 | 1.12 | 1.11 | 1.18 | 1.17 | 1.15 | 1.21 |
| Pb | 2.07 | 0.04 | 0.11 | 0.008 | 0.008 | 25.31 | 1.12 | 0.42 | 0.02 |
| Fe | 1.72 | 0.47 | 1.16 | 1.18 | 2.14 | 8.10 | 3.14 | 3.83 | 3.75 |
| Ni | 2.76 | 0.24 | 0.27 | 0.00 | 0.00 | — | — | — | — |
| Co | 1.14 | 0.21 | 0.16 | 0.00 | 0.00 | — | — | — | — |
| $Al_2O_3$ | — | 0.04 | 0.35 | 0.46 | 0.97 | — | 0.22 | 5.07 | 4.92 |
| $MnO_2$ | — | 0.00 | 0.00 | 0.09 | 0.22 | — | — | 0.20 | 0.30 |
| CaO | — | 0.00 | 0.15 | 0.18 | 0.17 | — | 0.22 | 0.22 | 0.06 |
| $P_2O_5$ | — | 0.53 | 2.33 | 1.98 | 1.96 | — | 3.41 | 5.41 | 5.40 |
| $SO_3$ | — | — | — | 0.22 | 0.09 | — | 0.45 | 0.11 | 0.04 |
| $SiO_2$ | 0.02 | 0.02 | 0.14 | 0.15 | 0.10 | — | 1.66 | 7.18 | 6.43 |
| $CO_2$ | — | 0.98 | 3.23 | 1.18 | 0.78 | — | 29.40 | 3.58 | 0.10 |
| Glühverl. | — | — | 37.83 | 34.04 | 36.94 | — | 28.34 | 43.73 | 42.79 |
| Humus | — | — | 0.37 | 0.89 | 0.91 | — | — | — | 0.16 |

nach Formen, die der Mensch anderem Material, nämlich Bronze, gegeben hatte." Später sind derartige Stücke noch mehrfach beschrieben und untersucht, 5) jedoch fehlten Präzisionsanalysen, bei denen alle vorkommenden Stoffe berücksichtigt sind. Dieser Mangel war zu beheben, als eine größere Zahl geeigneter Fundstücke zur Untersuchung zur Verfügung gestellt werden konnte, sodaß die Untersuchungsergebnisse für einige ausgewählte Proben hier angeführt werden können.

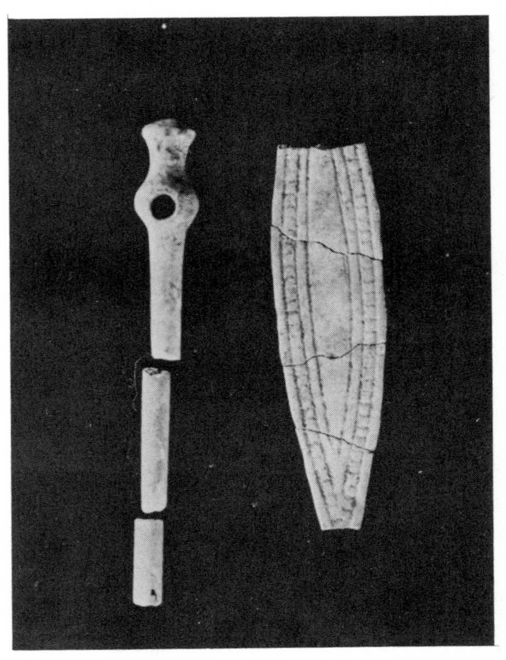

**2. Spangenfibel von Barchel. 1200 v. Chr. Ubergegangen in Zinnsäure. Nat. Gröβe.**

Probe 1. Spangenfibel. Der in Bruchstücken geborgene Fund kam bei Barchel, Krs. Bremervörde zu Tage und gehört dem mittleren Abschnitt der nordischen Bronzezeit an (1200 v. Chr.)

Die im Bild 2 wiedergegebenen Teile sind so vorzüglich erhalten, daß alle Einzelheiten der ursprünglichen Bronze erkennbar sind. Das Stück besteht aus einer gelblich-weißen, heute sehr festen Masse, die nur stellenweise einige gelbliche oder grünliche Flecke aufweist.

Probe 2. Reste eines stark profilierten Reifes, der in einem zerfallenen Holzgefäß in einem Hügelgrabe bei Harmelingen, Krs. Soltau, 1853 gefunden wurde.

Die ursprüngliche Form ist sehr gut erhalten, sodaß die eingepunzten Vertiefungen und das flache Profil noch gut zu erkennen sind. Vergl. Bild 3. Die Masse selbst ist relativ fest, fast weiß gefärbt und zeigt nur wenige bräunlich oder grünliche Flecke.

**3. Profilierter Armreif. Ubergegangen in Zinnsäure; 1.25 x vergr.**

Probe 3. Radnadel, einem in einem Hügelgrab bei Reessum, Krs. Rotenburg gefundenem eiförmigem Tongefäß entnommen.

**4. Radnadel von Reessum, mittlere Bronzezeit, restlos in Zinnsäure umgewandelt; 2 x vergr.**

Die in Stücke zerbrochene, aber sonst vorzüglich erhaltene Nadel ist 22.2cm lang, mit einem Nadelkopf von 5.0 cm Durchmesser. Die Oberfläche ist völlig glatt, die Masse rein weiß mit wenigen bräunlichen und grünlichen Flecken. Bild 4.

Probe 4. Armreif ebenfalls aus Reessum und mit der Radnadel geborgen.

Ein Armreif von nahezu kreisförmigem Querschnitt (Stärke etwa 7mm der im Aussehen an schwach braunen, gebrannten Ton mit glasierter Oberfläche erinnert. Im Querschnitt fast weiß mit einer scharf begrenzten braungrünen Zone, (vergl. Bild 5) in der etwas reichlicher Kupferkarbonat und Eisenoxyd neben Humus nachzuweisen ist, während die schwache Färbung im Innern allein durch Humuseinlagerungen erzeugt ist.

Probe. 5 Reste einer kleinen Dolchklinge, wahrscheinlich

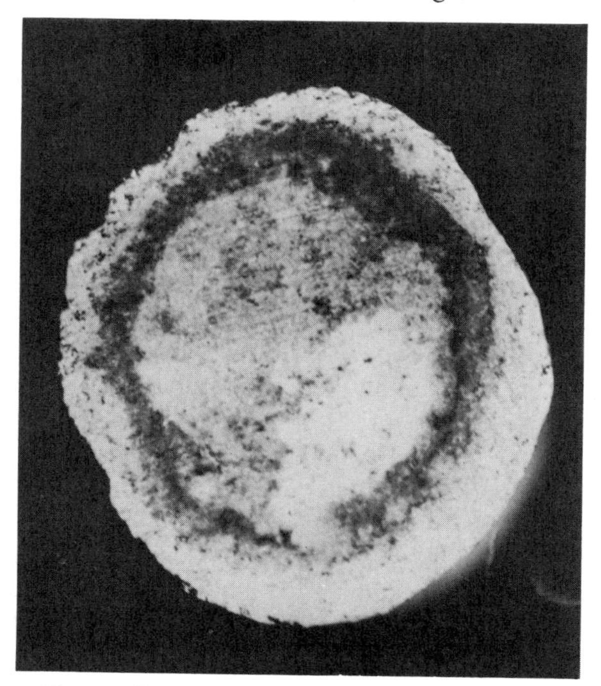

**5. Armreif von Reessum, mittlere Bronzezeit Querschnitt mit orientur eingelagerter Eisen-und Humuszone; 11 x vergr.**

aus einem Hügelgrab der Ülzener Gegend, Zeit etwa 1200 v. Chr.

Die 2–3cm breite, in der Mitte 3–4mm starke Klinge ist von schwach bräunlicher Farbe und oberflächlich so gut erhalten, daß die alten Kratzer und Schliffspuren noch zu erkennen sind.

Die gefundene prozentische Zusammensetzung der aufgeführten Proben enthält die nachfolgende Tabelle.

Wie zu erwarten war, besteht der Verwitterungsrückstand im wesentlichen aus Zinndioxydhydrat, das nur noch geringe Mengen an Kupfer führt, die bis zu 0,7% abfallen. Das aus dem Metall herrührende Antimon und Arsen sind als Sauerstoffverbindungen beim Zinn verblieben, während Blei, Nickel und Kobalt restlos verschwunden sind. Die relativ hohen Gehalte an Kohlendioxyd sollten einem noch vorhandenem Gehalt an Kupferkarbonat entsprechen, jedoch ist die Menge an $CO_2$ bei weitem nicht ausreichend, um das vorhandene CuO als Karbonat zu binden, sodaß ein Teil in anderer weniger gefärbter Verbindung vorliegen muß, vielleicht auch vom Zinndioxyd adsorptiv gebunden ist. Überraschend ist der zum Teil hohe Gehalt an Phosphorsäure, die von außen zugeführt und wahrscheinlich durch Zinnsäure adsorptiv gebunden ist. Die sonstigen geringen Mengen an Begleitstoffen müssen aus dem Bodenwasser stammen, sie sind zu gering, um für irgend welche Mitwirkung bei der Verwitterung in Betracht zu kommen.

Die geringen Humusgehalte entstammen ebenfalls den Bodenlösungen mit denen sie in das Zinnsäuregel einwanderten und in bestimmten Zonen gemeinsam mit Eisen- und Manganhydroxyden zur Ausfällung kamen, wie die im Bilde 5 auftretende ringförmige, dunkelbraune Zone erweist.

Besonders frappierend ist die Tatsache, daß bei dieser doch recht weit gehenden Umwandlung, bei der rd. 90% des ursprünglichen Metalls verloren gehen, die alte Form mit überraschender Naturtreue erhalten bleibt, sodaß mit Recht von einer "Pseudomorphose von Zinnsäure nach Bronze" gesprochen werden kann.

TABELLE 5.

Prozentische Zusammensetzung der Verwitterungsreste von Bronzen

| | 1 % | 2 % | 3 % | 4 % | 5 % |
|---|---|---|---|---|---|
| $SiO_2$ | 0.15 | — | 0.11 | 2.20 | 0.32 |
| CuO | 0.69 | 5.92 | 2.14 | 2.89 | 0.45 |
| PbO | — | 0.34 | — | 0.08 | — |
| $SnO_2$ | 77.85 | 65.20 | 69.52 | 70.27 | 76.31 |
| $Sb_2O_5$ | 0.79 | 0.52 | 0.52 | 1.03 | 1.34 |
| $As_2O_5$ | 0.08 | 1.07 | Spur | 0.80 | 0.09 |
| $Fe_2O_3$ | 3.02 | 1.92 | 4.62 | 0.56 | 3.21 |
| $Al_2O_3$ | 0.98 | 0.20 | 0.20 | 0.10 | 0.80 |
| $MnO_2$ | Spur | 0.05 | 0.08 | 0.03 | 0.10 |
| NiO | — | Spur | — | — | — |
| CaO | 0.05 | 0.04 | — | Spur | — |
| $P_2O_5$ | 1.86 | 3.52 | 2.14 | 2.78 | 2.31 |
| $SO_3$ | — | 0.14 | 0.20 | Spur | — |
| $CO_2$ | 0.05 | 0.85 | — | 0.10 | 0.53 |
| Glühverl. | 15.13 | 20.25 | 20.62 | 19.59 | 14.65 |
| Summe | 99.95 | 100.02 | 100.15 | 100.43 | 100.11 |
| Humus | Spur | 0.08 | 0.10 | 0.28 | 0.36 |
| $NH_4$ | — | 0.03 | — | 0.03 | — |
| $H_2O$ | 15.10 | 19.29 | 20.32 | 19.18 | 14.29 |

Als völlig abwegig abzulehnen ist der Gedanke, daß es sich bei diesen Massen überhaupt nicht um die aus verwitterten Bronzen entstandene Zinnsäure handelt, sondern daß sie aus metallischem Zinn gebildet sei. Hiergegen spricht das gelegentlich im Innern noch vorhandene Metall, das einwandfrei als Bronze zu erweisen ist und die Tatsache, daß Stücke zur Untersuchung kamen, die alle

Übergänge vom Metall bis zur Zinnsäure aufweisen. Im übrigen dürfte der Übergang vom Zinn in wasserhaltige Zinnsäure mit einer derartigen Volumvergrößerung verknüpft sein, daß eine Erhaltung der ursprünglichen Form in einer derartigen Feinheit ausgeschlossen erscheint.

TABELLE 6.
Phosphathaltige Patina

| | *Spiralarmreif Vorwohlde* | | *Dolch Benefeld* | *Dolch Wellendorf* | *Dolch Lüneburg* |
|---|---|---|---|---|---|
| | *oben* | *unten* | | | |
| | 1 | 2 | 3 | 4 | 5 |
| CuO | 52.10 | 51.04 | 35.91 | 40.86 | 35.04 |
| $SnO_2$* | 19.91 | 14.76 | 6.09 | 0.52 | 3.63 |
| PbO | — | — | — | Spur | — |
| $Fe_2O_3$ | 2.18 | 1.72 | 1.07 | 2.70 | 4.75 |
| CaO | Spur | Spur | 0.53 | 5.58 | 5.97 |
| $P_2O_5$ | 1.46 | 5.86 | 21.37 | 22.62 | 25.85 |
| $SO_3$ | Spur | Spur | Spur | — | — |
| Cl | Spur | Spur | — | — | 0.05 |
| Glühverl. | 24.86 | 27.00 | 35.03 | 27.29 | 24.07 |
| Summe | 100.43 | 100.38 | 100.00 | 99.57 | 99.35 |
| Verhältnis Cu/Sn | 2.64:1 | 3.49:1 | 5.98:1 | 81.56:1 | 9.79:1 |

*Enthält Spuren $Sb_2O_5$ und $As_2O_5$ (bis zu 1/10%).

5. *Phosphathaltige Patina.* Geringe Phosphatgehalte sind in der Patina häufig zu beobachten. Sie sind nicht durch einen Phosphorgehalt der Bronze verursacht, sondern allein durch eine Einwanderung aus der Umgebung zu erklären, wo die Gelegenheit zur Bildung stärker phosphorhaltiger Lösungen durch gleichzeitig vorhandene und verwitternde Knochen fast immer gegeben war. Wie aus den in Tabelle 6 zusammengestellten Analysenwerten hervorgeht, können gelegentlich auffallend hohe Gehalte auftreten, die dann, wie später gezeigt wird, ein wertvolles Beweismittel für den Nachweis eines völlig vergangenen Materiales bilden können.

Der Spiralarmreif von Vorwohlde entstammt einer Erdbestattung und umschloß noch den Armknochen, von diesem getrennt durch zwischenliegende Erde. In dem oberhalb des Knochens liegenden Abschnitt des Reifes enthielt die Patina 1.5% $P_2O_5$, während unterhalb des Knochens und vor direkter Berührung mit ihm durch eine etwa 2cm starke Bodenschicht geschützt in der Patina 5.85% ermittelt wurden. Dieser wesentlich höhere Gehalt kann nur durch abwärts sickernde Bodenlösungen verursacht sein, die aus den verwitternden Knochen Phosphate lösten, die sich mit der Patina des tiefer liegenden Teiles des Armbandes umsetzten.

*6. Verbleib der aus der Bronze gelösten Metalle.* Die beim Verwitterungsvorgang aus der Bronze gelösten Metalle können vom Boden durch Adsorption oder chemisch gebunden festgelegt werden, wie durch Analyse der eine verwitternde Bronze umgebende Bodenmasse leicht festzustellen ist. Für einige humose Sandböden konnte die Bestätigung dieser Annahme erbracht werden, wie aus der Tabelle 7 hervorgeht.

Jeweils 10g Boden wurden durch mehrfaches Abrauchen mit Königswasser aufgeschlossen und in der nach Aufnahme mit 1 n. Salzsäure erhaltenen Lösung die Bestandteile ermittelt.

In allen Proben finden sich reichlich Kupfer und geringere Mengen Nickel und Blei, also die Metalle, die bei dem Abbau der Bronze verschwinden. Zinn, Antimon und Arsen treten nur in Spuren bezw. untergeordnetem Gehalte auf. Dagegen ist der Gehalt an löslicher Phosphorsäure, besonders in den Böden aus Gockenholz recht hoch, was nur durch Festlegung der aus den Knochen der Bestatteten löslich gewordenen Phosphate zu erklären ist. Die anderen

Zahlen geben einen Anhalt für die Menge der sonst noch im Boden vorhandenen Verbindungen, von denen einige in die hinterbliebene Zinnsäure einwandern.

TABELLE 7.

Metallgehalte in Bodenproben
in mg pro 10g

| | | *Gockenholz* | | |
|---|---|---|---|---|
| | *Bookholt* | *I* | *II* | *Vorwohlde* |
| CuO | 144.0 | 300.0 | 257.0 | 422.0 |
| PbO | 0.9 | 0.5 | 0.5 | — |
| NiO | 0.1 | 0.3 | 0.3 | 1.2 |
| $SnO_2$ | 0.1 | 0.8 | 0.3 | — |
| $As_2O_5$ | Spur | Spur | Spur | Spur |
| $Sb_2O_5$ | Spur | Spur | Spur | — |
| $Fe_2O_3$ | 22.4 | 51.7 | 58.3 | |
| $Al_2O_3$ | 20.6 | 24.4 | 46.3 | |
| MnO | 0.8 | 1.3 | 1.0 | |
| CaO | 2.2 | 5.6 | 5.3 | |
| MgO | 2.4 | 2.2 | 2.2 | |
| $P_2O_5$ | 5.8 | 96.4 | 90.2 | |
| $SO_3$ | 0.5 | 3.8 | 3.8 | |
| Cl | 0.5 | 1.0 | 0.8 | |
| Humus | 250.0 | 175.0 | 175.0 | |

Auch rein experimentell ist die Fähigkeit der Böden, Metallsalze aus Böden aufzunehmen und zu binden, leicht festzustellen. Wurden jeweils 10g der Sandböden von Gockenholz mit 100ml neutraler, wäßriger Metallsalzlösungen (angewandt wurden die Sulfate) steigender Konzentration 2 Stunden geschüttelt und anschließend das in Lösung verbliebene Metall ermittelt, so zeigte sich, daß bei geringen Konzentrationen das Metall restlos aus der Lösung verschwand. Bei größeren Gehalten wird unabhängig von der Konzentration nur ein bestimmter, aber konstanter Teil aufgenommen, der der Aufnahmekapazität des Bodens entspricht, die bei den beiden Böden für Kupfer bei 12 bezw. 18mg/10g Boden lag. Für Nickel, Blei und Zink gilt etwa das Gleiche.

Für andere Bodentypen gelten andere Zahlen, die aber oft höher liegen.

In welcher Form das Metall gebunden wird steht dahin. Außer der Bindung durch reine Adsorption dürfte auch eine Festlegung durch chemische Bindung erfolgen, sobald reaktionsfähige Stoffe im Boden vorhanden sind, z. B. Calciumkarbonat, die die Möglichkeit zur Bildung schwer- bezw. unlöslicher Metallverbindungen geben.

Eine chemische Bindung der aus der Bronze bezw. ihrer Patina gelösten Metallsalze ist auch durch die mit ihnen in den Gräbern oder Urnen in Berührung kommenden Knochen möglich, wo die Festlegung in der Form von Phosphaten erfolgte. Diese Tatsache ist für Kupfer allgemein bekannt, da das abgeschiedene Phosphat durch die Verfärbung des Knochens sofort auffällt, aber auch die sonstigen abwandernden Metalle werden gebunden, wie aus der Zusammenstellung in Tabelle 8 hervorgeht.

TABELLE 8.

Metallgehalt in org. Stoffen
mg Oxyd auf 10g Probe berechnet

| *Probe* | *CuO* | *PbO* | *NiO* | *ZnO* |
|---|---|---|---|---|
| 1. Leichenbrand | 122.0 | 1.1 | — | 8.0 |
| 2. Leichenbrand | 255.5 | 20.3 | — | — |
| 3. Knochenrest | 3916.0 | Sp. | 0.53 | — |
| 4. Knochenrest | 265.0 | Sp. | Sp. | — |
| 5. Rez. Knochen | 427.0 | — | — | — |
| 6. Holz | 2904.0 | 0.15 | 0.62 | — |
| 7. Holz | 285.0 | Sp. | Sp. | 30.0 |
| 8. Geweberest | 1485.0 | Sp. | | 5.0 |
| 9. Geweberest | 2250.0 | Sp. | | 7.0 |

1. Leichenbrand aus einem römischen Bronzeeimer von Hemmor des 2.–3. Jahrh. n. Chr. aus stark zinkhaltiger Bronze. Die Knochenreste sind soweit sie mit dem Metall in Berührung kamen, bräunlich grün verfärbt.

2. Leichenbrand aus der Bronzeurne von Helzendorf (siehe

dort entnommen, wo die Verwandlung in Zinnsäure bereits weitgehend fortgeschritten ist. Der Knochenrest ist bräunlich grün verfärbt.

3. Knochenrest aus dem bereits erwähnten Spiralarmband von Verwohlde. Die obere Seite ist stärker bräunlich grün verfärbt als die untere Seite, bedingt durch absickernde Kupferlösung. Hier konnte röntgenographisch der Nachweis von Libethenit geführt werden. 1)

4. Bräunlich grün verfärbter Knochenrest aus einem Hügelgrabe, der unmittelbar neben einer Bronze lag. (Speerspitze.)

5. Rezenter Knochen, oberflächlich braun grün gefärbt, der etwa 30Jahre in 2–3cm Abstand unter einem als Blitzableiter Elektrode in 50cm Tiefe verlegtem 3cm breitem Bande aus Reinkupfer lag. Das Kupfer war bereits stark patiniert.

6. Bräunlich verfärbtes Holz aus der Tülle einer Speerspitze (etwa 1000 v. Chr.)

7. Holzstück aus einem römischen, stark Zink-haltigem Bronzegfäß des 2–3.Jahr. n. Chr.

8 und 9. Völlig mit Patina inkrustierter Geweberest aus römischen, zinkhaltigen Bronzegefäßen. 6)

Die aufgenommenen Metallmengen können, wie die Zahlen der Tabelle ausweisen, recht beträchtlich werden. In keiner der Proben ist Zinn, Arsen oder Antimon nachweisbar, ein Beweis dafür, daß sie beim Ablauf der Verwitterung nicht löslich werden.

In den Knochenresten erfolgt stets die Festlegung als Phosphat, als Libethenit, wie röntgenographische Untersuchungen ergaben. 1) Im Holz und den Geweben liegen Karbonate vor.

*7. Die Ursachen der Metallösung aus Bronzen.* Für die

Fortlösung des Kupfers aus Bronzen und ihrer Patina können eine größere Zahl von Faktoren in Betracht kommen. Die im täglichen Leben beobachteten Ursachen für die Korrosion von Kupfer und Kupfer-legierungen, wie die Einwirkung stärkerer anorganischer oder organischer Säurelösungen oder wäßrige Salzlösungen hoher Konzentration, elektrolytische Prozesse und andere mehr, sind von vornherein auszuschließen, da sie natürlicherweise in den humosen Sandböden nicht auftreten. Die hier zu erwartenden Konzentrationen sind viel zu gering um selbst bei langfristiger Einwirkung in solchem Maße in Erscheinung treten zu können.

Das Gleiche gilt von den bei der Verwesung eines Bestatteten auftretenden Produkten mit ihrem Gehalt an Ammonsalzen oder freiem Ammoniak, die man mehrfach verantwortlich gemacht hat. 7). 8). Hiergegen spricht die Beobachtung, daß die Zersetzung nicht auf Erdbestattungen beschränkt ist, sondern in gleicher Weise auch bei den Bronzen aus Brandgräbern und Urnen mit Leichenbrand beobachtet wird, wo keine Fäulnisvorgänge abgelaufen sind.

Weiter ist zu beachten, daß die Fäulnis ein anaerober Vorgang ist, bei dem organische Schwefelverbindungen und Schwefelwasserstoff entstehen, die zur Bildung von Metallsulfiden führen, die in alkalischen Lösungen der zu erwartenden Konzentration völlig unlöslich sind. Im übrigen ist die absolute Menge der nur eine relativ kurze Zeit vorhandenen Fäulnisprodukte so gering, daß die beobachteten weitgehenden Zersetzungen des Metalles gänzlich ausgeschlossen sind. Eine Löslichkeit von Kupfer in Ammoniak und Ammonsalzlösungen verlangt, abgesehen von der in einem Grabe nie zu erreichenden Konzentration, stets die Gegenwart von Sauerstoff, der aber während der Möglichkeit einer Ammoniakbildung im Grabe völlig fehlt. Fäulnisvorgänge sind für die Fortlösung des Kupfers aus Bronzen daher keinesfalls verantwortlich, worauf auch schon frühzeitig hingewiesen wurde. 1)

Wie bereits Ausgang des 19.Jahrhunderts klar erkannt und ausgesprochen wurde, 7–9 und heute wohl allgemein anerkannt ist, kommt sowohl für die Bildung der Patina als auch ihre Zersetzung allein Wasser mit gelöstem Sauerstoff und Kohlendioxyd in Betracht. Allen anderen möglichen Faktoren im Boden ist nur eine gelegentliche und untergeordnete Mitwirkung zuzusprechen.

In den wasser-und luftdurchlässigen humosen Heidesanden mit ihrer eignen hohen Kohlendioxyproduktion sind die Grundlagen für eine bis zur reinen Zinnsäure verlaufenden Zersetzung der Bronzen eher gegeben, als in schwerer durchlässigen oder Kalk führenden Erdschichten.

Experimentelle Untersuchungen über die Löslichkeit von basischem Kupferkarbonat, dem Hauptbestandteil der Patina, in Kohlendioxyd-haltigem Wasser sind mehrfach durchgeführt.

Für die Beurteilung der hier interessierenden Fragen scheint die Untersuchung von E. E. Free 10) von besonderer Bedeutung. Hier wurde gezeigt, daß die Löslichkeit allein von der Konzentration an freiem Kohlendioxyd im Wasser abhängt und daß sie durch die Gegenwart geringer Neutralsalzgehalte wie Natriumchlorid, Natriumsulfat oder Calciumsulfat, wie sie normalerweise im natürlichen Wasser auftreten, nicht beeinflußt werden. Erst Gehalte von 2g pro Liter erhöhen die Löslichkeit beträchtlich.

Eine ausnahme macht die Gegenwart von Natriumkarbonat und Calcium-hydrogenkarbonat, die die gelöste Kupfermenge erheblich herabsetzt. Löst beispielweise 1 Liter mit $CO_2$ gesättigtes Wasser 35mg Kupfer, so geht diese Menge bei Gegenwart von 100mg $CaCO_3$ und gleichem $CO_2$ Gehalt auf etwa 7mg zurück und bereits 10mg $CaCO_3$/1L Wasser genügen, um bei gleichen Versuchsbedingungen die Cu-menge von 35 auf 25mg zu erniedrigen.

Da die in den aus humosen Sandböden austretenden wasser sehr arm an gelöstem Calcium sind, kommt für sie eine merkliche Erniedrigung für die Löslichkeit der Patina kaum in Betracht, im Gegensatz zu stärker kalkhaltigem Boden, wo die vorzügliche Patinierung der Bronzen nicht zerstört wird.

Untersuchungen an Wasserproben aus humosen Heidesandgebieten zeigten, daß der Wasserstoffexponent der frischen Wasserproben sich zwischen 3.8 und 6.3 bewegt, der Gehalt an freiem $CO_2$ zwischen 25 und 100mg/1L. schwankt und zum Teil abhängig von der Jahreszeit der Probenahme ist.

Systematische Versuche, die mit einer Aufschlämmung eines Gemisches von Kupferkarbonat und Zinnsäure * durchgeführt wurden zeigten, daß beim Schütteln mit $CO_2$-haltigem Wasser schnell Kupfer, aber niemals Zinn in Lösung ging und die gelöste Kupfermenge mit zunehmendem $CO_2$ Gehalt anstieg.

Wurde mit $CO_2$ gesättigtem Wasser gearbeitet, und dieses wiederholt auf den nach Abtrennung der Lösung jeweils verbleibenden Bodenkörper einwirken lassen, so konnte das Kupfer weitgehend herausgelöst werden, während Zinn restlos zurückblieb.

Die in gleicher Weise gemeinsam mit Zinnsäure aus-

* Für die Versuche wurden jeweils 0,5g eines Gemisches von Kupferkarbonat und Zinnsäure verwandt, das folgenderweise erhalten war. Eine Auflösung von 9g Kupfer und 1g Zinn in wenig Königswasser wurden nach Verdünnung mit dest. Wasser auf 2 Liter durch eine hinreichende Menge an Natriumkarbonat gefällt. Anschließend wurde filtriert und Chlorid frei gewaschen.

gefällten Karbonate von Blei, Nickel und Zink verhielten sich dem Kupfer analog, auch sie wurden durch $CO_2$ haltiges Wasser mehr oder weniger schnell aus dem unlöslichen Zinnsäuregel gelöst.

Wurde die künstliche Patina dagegen mit rein wäßrigen oder $CO_2$-haltigen Lösungen mit geringen Gehalten an Eisen (III) sulfat oder Kaliumaluminiumsulfat geschüttelt, so wurde wiederum Kupfer gelöst, während Eisen und Aluminum quantitativ in den Bodenkörper übergingen, was nach den für die Fällung beider gültigen Wasserstoffexponenten auch zu erwarten war. Diese Umsetzungen erfolgen quantitativ in wenigen Minuten noch mit Lösungen, die pro Liter mg und weniger an ihnen enthalten.

Beim Schütteln der "künstlichen Patina" mit Kohlendioxyd und Phosphorsäure führendem Wasser wird letztere vollständig vom Bodenkörper aufgenommen, während wie üblich, Kupfer entsprechend der $CO_2$ Konzentration gelöst wird. Die Fällung der Phosphorsäure ist auch dann vollständig, wenn Milligrammengen pro Liter zur Anwendung kamen. Ob die Festlegung durch chemische Bindung an Kupfer oder durch Adsorption an Zinnsäure erfolgte, steht dahin und dürfte für unsere Betrachtungen zunächst belanglos sein.

*8. Verlauf der Patinabildung und der ihrer Zerstörung.* Nach den mitgeteilten Untersuchungsbefunden, ergänzt durch mikroskopische Untersuchung von Anschliffen, dürfte sich ein relativ zuverlässiges Bild von der Bildung der Patina und ihrer Zerstörung sowie der Entstehung der "Pseudomorphosen von Zinnsäure nach Bronzen" ergeben.

Zunächst erfolgt, von der Oberfläche ausgehend und an den Korngrenzen des Gefüges nach dem Innern zu fortschreitend, eine Oxydation des Metalles. Bild 6 zeigt bei

100facher Vergrößerung das Bild an der Grenze der Oxydationszone einer Bronze, die beim Anätzen die typische Gußstruktur hervortreten läßt, ungeätzt aber gleichmäßig hell erscheint, wie im linken Teil des Bildes.

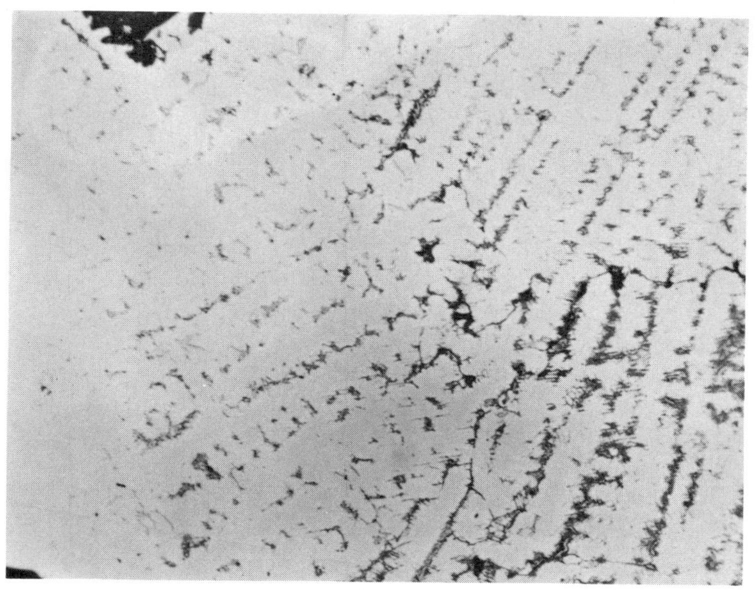

**6. Kupferoxydul auf Korngrenzen in der Bronze. Anfangsstadium Anschlift im auffallenden Licht; Vergr. 100 x fach.**

Das an den Korngrenzen auftretende $Cu_2O$ ist mikroskopisch an seiner typischen Färbung sicher zu erkennen, im Bilde erscheint es dunkel. Es tritt an den Gefügegrenzen auf und seine Menge nimmt vom Metall an zu. Bei fortschreitender Oxydation verbreitert sich das eingelagerte $Cu_2O$ (Bild 7), das sich in weiterem Abstande vom Metall, schnell ausbreitet und das $Cu_2O$ verdrängt, (Bild 8) wie die tief schwarzen Stellen des Bildes ausweisen. Schließlich liegt ein Gemisch von $Cu_2O$ und CuO vor (Bild 9) in dem das noch vorhandene $Cu_2O$ hell hervortritt, seine scharfen Umrisse aber bereits verloren hat.

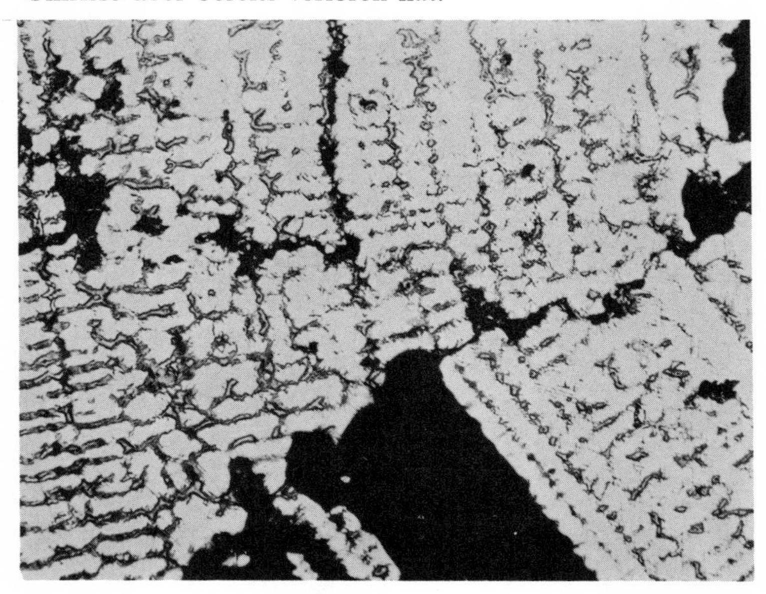

**7. Fortschritt der $Cu_2O$ Bildung, am unteren Rande schwarzes CuO. Anschliff-Auffallendes Licht; Vergr. 100 fach.**

Gelegentlich liegen im Oxydgemisch noch Metallreste, die im Bilde 10 hell hervortreten und völlig von den durch ihre unterschiedliche Farbe erkennbaren Oxyde umschlossen werden.

Patina oder Zinndioxyd sind nicht zu erkennen. Die blaugrüne Patinaschicht entsteht erst an der Grenze der Oxyde, in die sie sich zackig hereinfrißt. Auch in ihr ist kein $SnO_2$ sichtbar, das sich jedoch durch Aufhellung der Färbung kenntlich macht.

**8. Reichlich $Cu_2O$ neben schwarzem CuO. Fortschritt von Bild 7; Anschliff Vergr. 100 fach.**

Die Umwandlung des Metalles in das Oxydgemisch ist mit einem Kupferverlust verknüpft, wie sich in der Verschiebung des Kupfer-Zinnquotienten zu Gunsten des Zinns bei den Analysen der Oxydschichten ausprägt. (Vergl. Tab. 1.)

An die Oxydschicht schließt sich die mehr oder weniger blaugrüne Patinaschicht an, wobei der gegenseitige Übergang nicht scharf abgesetzt sondern allmählig erfolgt und gegenseitige Einfressungen in einander auftreten.

Das zunächst bei der Oxydbildung löslich gewordene Kupfer wandert an die Oberfläche und scheidet sich hier als Patinaschicht; ab, die wenig Zinn enthält, ja sogar davon frei sein kann, wie die Analyse sehr dünner, fest haftender Patinaschichten ausweist, bei denen der Kupfer-Zinnquotient wesentlich nach der Kupferseite verschoben ist (Tab. 1 and 2.)

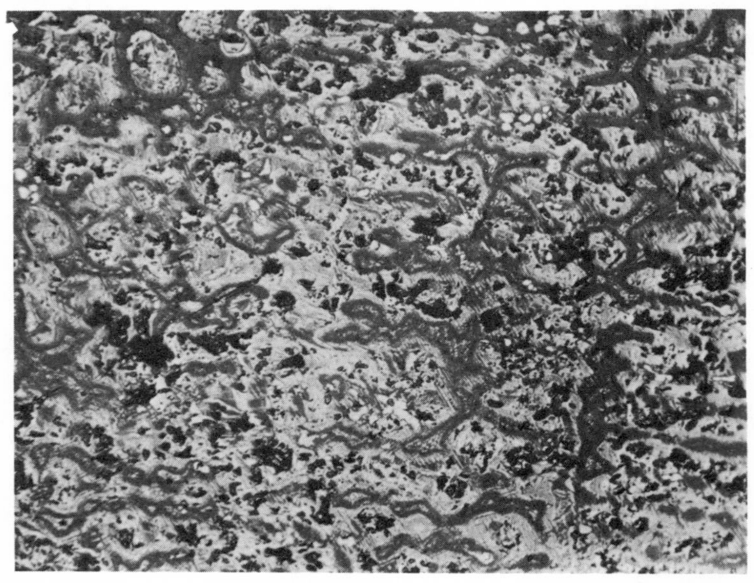

**9. CuO neben $Cu_2O$. Fortschritt von Bild 8; Anschliff Vergr. 100 fach.**

Bei weiterer Einwirkung eines Wassers mit hohem Kohlendioxydgehalt und mäßigem Sauerstoffgehalt, schreitet die Oxydation des Metalles fort, gleichzeitig macht sich aber auch die lösende Wirkung der $CO_2$ auf die Patinaschicht geltend, Kupfer wird gelöst und im Boden oder sonstwie festgelegt, während das Zinn, da nicht löslich, sich in der restlichen Patina anreichert, was sich in einer Verschiebung des Cu/Sn Quotienten zur Zinnseite hin ausprägt. Die Verarmung an Kupfer in der Patina schreitet nach dem Oxydkern vor, der gleichzeitig mehr und mehr in Karbonat umgewandelt wird, das weiter der Zersetzung unterliegt. Mit fortschreitendem Kupferverlust reichert sich die unter den Verwitterungsbedingungen unlösliche Zinnsäure an und hinterbleibt schließlich allein. Mit den Bodenlösungen sind Eisen, Phosphat und Humus dem Zinnsäuregel zugeführt, und in ihm durch Adsorption oder Ausfällung, oft in orientierten Fällungen, abgeschieden ebenso wie das in der Bronze vorhandene Antimon und Arsen, die als Oxyde von Anfang an von der Zinnsäure festgehalten wurden, während die 2 wertigen Elemente wie Blei, Nickel, Kobalt und Zink sich wie Kupfer verhielten und der Auswaschung anheimfielen.

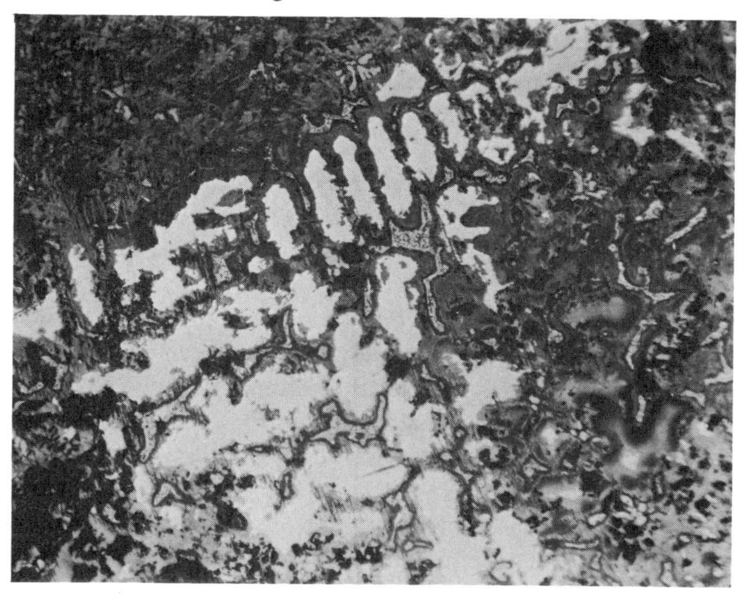

**10. Bronzereste (hell) im $Cu_2O$-CuO Gemisch; Anschliff Vergr. 100 fach.**

Wirkte reichlich Wasser mit besonders hohem $CO_2$ und $O_2$ Gehalt auf Bronzen ein, wie es in den leicht durchlässigen humosen Sanden der Fall ist, so konnte hier die völlige Ausbildung von "Pseudomorphosen von Zinnsäure nach Bronzen" erfolgen, die überraschender Weise alle Einzelheiten des ursprünglichen Metalles wiedergeben.

Wurde jedoch, etwa durch teilweise Überdeckung mit Keramikscherben, der Wasserzutritt beschränkt, so wurde hier der Abbau verzögert, was dazu führte, daß am gleichen Stück alle Stadien der Zersetzung vom Metall bis zur Zinnsäure beobachtet sind, wie es beim Schwert von Wölpe beschrieben ist.

Fehlt es im Wasser an $CO_2$, so tritt die Patinabildung und ihre Zersetzung zu Gunsten der Oxydbildung zurück und die Bronze ist weitgehend unter einer verhältnismäßig dünnen, vor allem zinnarmer Patinaschicht in ein Gemisch der Oxyde verwandelt.

Fehlt es am Sauerstoff, wie es in reinem Moorwasser der Fall ist, so tritt trotz erheblichem $CO_2$ Gehalt keine Karbonatpatina und tiefer gehende Zersetzung der Bronze ein. Unter einer dünnen braunen "Moorpatina" erscheint das intakte Metall.

In weniger durchlässigen Tonböden und Lehmböden sind infolge Hemmung Patinabildung und Zersetzung meist weniger stark ausgebildet. Das Gleiche gilt für Böden mit einem mehr oder weniger großem Kalkgehalt, wo der Gehalt an Calciumhydrogenkarbonat die Kupferlösung aus der Patina zurückdrängt.

Völlig andere Verhältnisse der Patinabildung sind bei einer Mitwirkung mehr oder weniger hoher Chloridgehalte im Wasser gegeben.

Grundsätzlich dürften die für Sandböden betrachteten Vorgänge allgemeine Gültigkeit haben, jedoch kann der Ablauf durch alle möglichen Variationen der für die

Patinabildung und Zersetzung maßgebenden Faktoren auch für gleiche Bodentypen zu unterschiedlichen Bildern führen.

In allen Fällen ist in der einer Bronze aufliegenden Patina eine Verschiebung der Menge der Bronzebestandteile im Vergleich zu der im Metall eingetreten, sodaß ein Rückschluß aus der Zusammensetzung der Patina auf die des Metalls völlig unmöglich ist.

b. Untersuchungen an vorgeschichtlichen Bronzewaffen.

Die durch Bronzeniete gehaltenen Griffe sind, da meistens organische Naturstoffe, beim Lagern im Boden fast immer so weit vergangen daß, nicht mehr zu erkennen ist, aus welchen Stoffen die Griffe ursprünglich gemacht waren. Unter besonders günstigen Bedingungen sind noch geringe Reste, vor allem um und zwischen den Nieten erhalten oder die Markierungen der alten Griffgrenzen sind noch erkennbar, wie auf den Bildern 11 und 12.

Bild 11 zeigt den Kopfteil mit Nieten eines bei Anderlingen, Krs. Bremervörde gefundenen Dolches aus der Zeit 1200 v. Chr. Die Patina läßt einen halbkreisförmigen Heftabschluß mit seitlichen Ecken erkennen. Zwischen den Nieten findet sich eine dicke, bräunlichgrüne Patinaschicht, die einen blättrigen Aufbau zeigt und als ein mit Kupfersalzen imprägnierter Rest des Heftes angesprochen wurde.

Bild 12 gibt das Aussehen des Griffteiles eines Griffzungenschwertes vom Unterelbetypus wieder, wo auf der Zunge und um die Niete eine 2mm starke, weißlich grüne Patinaschicht liegt, die sich leicht abheben und entfernen läßt, die aber die Grenzlinien zwischen Griff und Klinge deutlich ausprägt.

Auf Grund des Aussehens dieser Reste ist eine sichere

Angabe des ursprünglichen Griffmaterials nicht möglich, das gewesen sein kann Holz der verschiedensten Art, Knochen, Elfenbein, oder Hirschgeweih, um nur einige Möglichkeiten anzudeuten.

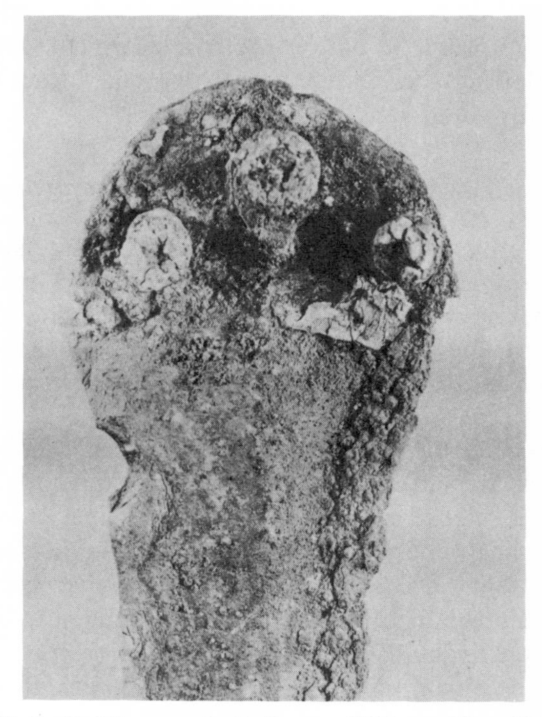

**11. Griffteil, Dolch von Anderlingen 1200 v. Chr. Neben den Nieten Reste der Griffmasse; 1.5 x vergr.**

Nach den bei der Patinauntersuchung gemachten Erfahrungen über das Verhalten von Phosphorsäure zur Patina und das von gelösten Kupfersalzen zu Knochen, hätte ein hoher Phosphatgehalt der Patina des Griffteiles für die Verwendung eines Phosphat-führenden Materiales gesprochen, sobald gleichzeitig gezeigt werden konnte, daß andere Bronzen keine oder nur beschränkte Mengen an Phosphorsäure aus dem Boden aufgenommen hatten, oder wenn der Phosphatgehalt der Patina am Griffteil wesentlich höher sein sollte als etwa an der eigentlichen Klinge, besonders zur Klingenspitze hin.

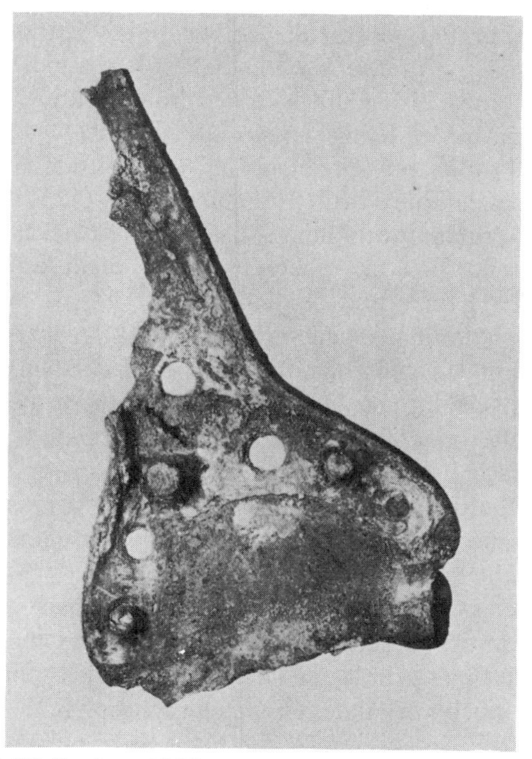

**12. Griffteil eines Griffzungenschwertes etwa nat. Größe.**

Die Untersuchung der Patinaproben einer größeren Zahl von Waffen ergab ein recht interessantes Bild.

*1. Patina von Beilen und Speerspitzen.* Die Patina von

Beilen, Speerspitzen und Armreifen führte fast immer geringe Gehalte an Phosphorsäure. Die Menge schwankte und bewegte sich zwischen Spuren und 1.4%, bezogen auf die an Sand freie Patinamasse. Bis zu 1.4% dürften demnach aus den Bodenlösungen in die Patina übergetreten sein.

*2. Patina von Dolchen und Kurzschwertern.* Bei der Prüfung der Patina von 8 Dolchen bezw. Kurzschwertern fanden sich um und in der Nähe der Niete 4 mal Phosphatgehalte, (etwa 20%) die 4–7 mal höher waren, als die in der Patina in der Klinge bezw. Klingenspitze.

Zwei Proben zeigten sowohl im Griffteil als auch in der Klingenspitze einen $P_2O_5$ Gehalt von 2.5–0.25%.

Bei 2 Proben mit hohem $P_2O_5$ Gehalt (20–22%) an den Nieten fand sich ein gleich großer Gehalt auch in der Patina der Klinge und der Klingenspitze.

In allen Patinaproben mit hohen $P_2O_5$ gehalten wurde gleichzeitig ein zwischen 3 und 5% liegender Gehalt an CaO festgestellt, der in den beiden $P_2O_5$ armen Proben ganz fehlte oder einige 1/10% ausmachte.

Demnach dürfte für zwei Dolche die Verwendung eines phosphathaltigen Naturproduktes für die Verwendung als Heft auszuschließen sein, sodaß hier wahrscheinlich Holz in Betracht kommt.

In vielen Fällen sind Knochen, Elfenbein oder Hirschhorn mit Sicherheit angewandt, wie nicht nur aus dem hohen Phosphatgehalt der Patina hervorgeht, sondern auch aus dem noch vorhandenen Calciumphosphat.

In zwei Fällen zeigen Griffteil und Klinge einen gleichen, aber recht hohen Phosphatgehalt und merklichen Kalkgehalt, ein Befund, der sofort erklärt ist, wenn man annimmt, daß die Scheide der Waffe ebenfalls aus Knochen gefertigt wurde, eine Vermutung, die durch andere Fundstücke belegt ist, auf denen Reste einer Knochenscheide beobachtet sind.

3. *Griffzungenschwerter.* Bei der Untersuchung einiger einfacher Griffzungenschwerter wurden ganz analoge Ergebnisse erhalten. Bei diesen wurde einerseits die von Griffzunge und Heftplatte entfernte Patina verwandt und anderseits die von der Klinge möglichst nahe der Spitze entnommene.

Bei 6 geprüften Schwertern zeigte nur eins für Griff und Klingenteil den nahezu gleichen geringen Gehalt von etwa 1.5% $P_2O_5$ und 1% CaO, was keinesfalls für eine Verwendung von Phophat-führendem Material spricht.

Bei 4 Stücken fanden sich im Griffteil $P_2O_5$ Gehalte von 10–17% und 2–5% CaO, während in den Klingenspitzen die Gehalte auf rd. 2% $P_2O_5$ neben 0.1–0.5% CaO abfielen.

Bei einer Klinge führte die Patina des Griffteiles 28% $P_2O_5$ neben 5% CaO gegen 18% $P_2O_5$ und 3% CaO in der Patina der nicht allzuweit von der Heftplatte abgebrochenen Klinge.

Bei einer Klinge ließ sich dicht an einem Niet von der Heftplatte ein Patinastückchen abheben, das auf der der Bronze zugewandten Seite bei schräger Beleuchtung eine Streifung erkennen ließ, die Bild 13 bei 14 facher Vergrößerung zeigt und aas weitgehend dem einer Spongiazone eines Längsschnittes einer Geweihstange ähnelt.

Diese zufällige Beobachtung ergänzt den chemischen Befund und erweist, daß Platten aus Geweihstangen für die Griffschale der Griffzungenschwerter verwandt sind, ein Brauch, der noch gültig ist, denn Griffe von Hieb-und Stichwaffen werden auch heute noch gern aus Abschnitten von Geweihstangen verfertigt.

4. *Patina auf Vollgriffdolchen und Vollgriffschwertern.*

Die Vollgriffwaffen entwickeln sich in der II. Periode der Bronzezeit, also zwischen 1550–1300 v. Chr. Technisch betrachtet sind es Griffangelwaffen mit einem Metallgriff. Diese Griffe sind durch eingeschlagene Muster verziert und zeigen oft größere Vertiefungen die mit reichlich bräunlichgrüner Patina gefüllt sind, deren Untersuchung einen überraschend hohen Gehalt an $P_2O_5$ und CaO ergibt, womit der Beweis erbracht ist, daß in diesen Vertiefungen Einlagen von Knochen oder Elfenbein angebracht waren.

**13. Patinastückchen mit Struktur; Vergr. 14 x, schräg auffallendes Licht.**

Als Beispiel für diese Gruppen sei zunächst der bei Westerwanna, Krs. Hadeln gefundene und im Landesmuseum Hannover aufbewahrte Vollgriffdolch erwähnt.

Er ist im Bilde 14 im Originalzustande abgebildet. Der einschließlich Knaufplatte 7cm lange Griff ist kreisrund und mit einer Knauflatte von 4cm Durchmesser und flachem Mittelkopf abgeschlossen, deren Rand durch 2 Zonen kurzer und dicker Kerben verziert ist.

**14. Vollgriffdolch von Westerwanna, Originalzustand.**

Auf Vorder-und Rückseite des Griffes ist in der Mitte eine Reihe konzentrischer Kreise eingepunzt, die spiralig durch schmale Brücken verbunden sind. Nach beiden Seiten schließen sich dreieckige Gruben an, die zur Aufnahme von Einlagen dienten. Der übrige Teil des Griffes ist durch mehrere in der Längsrichtung verlaufende Metallstege geteilt, deren Zwischenraum zickzackförmig verlaufende Stege zeigt.

Die Schultern des Heftes sind glockenförmig gewölbt, der Heftausschnitt ist dreiviertel kreisförmig. Das Heft trägt 4 Niete, von denen die beiden mittelsten jedoch nicht durch die Klinge hindurchgehen, sondern als Schmuckniete gelten können. Auf jeder Seite besteht jeweils 1 Niet aus Kupfer und einer aus Bernstein. Der Querschnitt des Dolchblattes zeigt eine breite Mittelrippe.

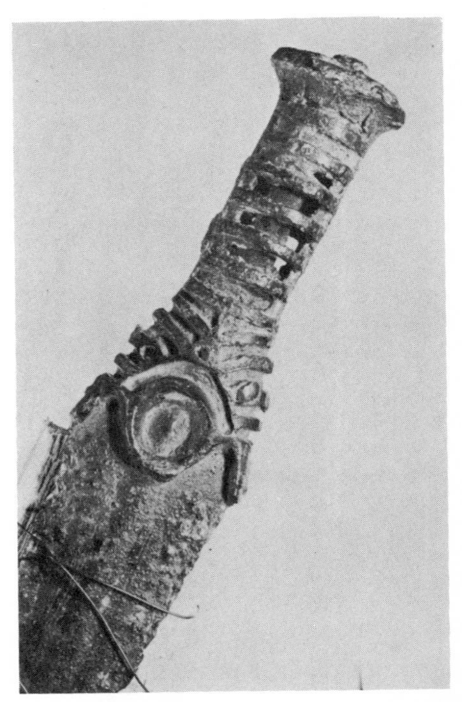

**15. Griffteil eines Kurzschwertes, Originalzustand.**

Sowohl die aus den Längsrippen als auch die aus den Dreiecken an den Kreisflächen entnommene Patina wies eine $P_2O_5$ von 6–8% auf, während eine Patinaprobe von der Klinge nur 3% und von der alten Klingenbruchfläche nur eine mikrochemisch nachweisbare Spur führte. Das Gleiche gilt vom CaO Gehalt, der in der Patina aus den Vertiefungen des Griffes zwischen 3 und 4% lag, auf der Klinge auf 0,1% abfiel und auf der Bruchfläche fehlte.

Ganz gleiche Beobachtungen wurden an dem im Bilde 15 wiedergegebenen Griffteil des vollständig erhaltenen Kurzschwertes gemacht. In der aus den tieferen Rillen entnommenen Patina fanden sich $P_2O_5$ Gehalte von 10–18% neben 3–5% CaO, während der $P_2O_5$ Gehalt auf der Klinge bis zu 1.5–2% absank,

Nach den vorstehend mitgeteilten Untersuchungsbefunden, die durch eine Anzahl weiterer Ergebnisse bestätigt und verbreitert sind, liegen die Unterlagen für eine materialgerechte Rekonstruktion vor.

Bild 16 zeigt die Rekonstruktion eines Dolches, bei dem als Griff der Abschnitt eines Stange des Rothirsches aufgesetzt und durch zwei Bronzeniete befestigt ist. Wie sich zeigte, liegt eine solche Waffe vorzüglich in der Hand, besonders da die geperlte Oberfläche des Griffes ein Rutschen verhindert.

Die Bilder 17 und 18 zeigen die im Niedersächsischen Landesmuseum angefertigten Nachbildungen des im Originalaussehen in den Photographien 14 und 15 wiedergegeben Dolches und Kurzschwertes. In die gegossenen Bronzeteile sind zugeschnittene Knochen bezw. Elfenbeinstücke eingepaßt, abgeschliffen und gemeinsam mit dem Metall poliert. Erst diese Nachbildungen vermitteln einen vollständigen Eindruck von der Schönheit einer solchen alten Waffe, den der zersetzte und patinierte Originalfund niemals ergeben kann und sie zeigen, daß nicht nur im Orient sondern auch im westeuropäischen Gebiet bereits im 2. Jahrtausend v. Chr. hochwertige Kunstwerke geschaffen wurden.

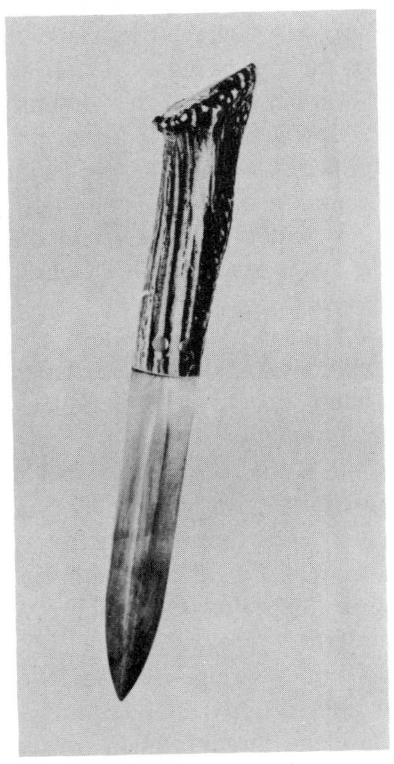

**16. Nachbildung eines Dolches mit Rothirschstange als Griff.**

5. *Griffzungenschwerter mit bleihaltigem Griff.* Als typischer Vertreter "der alten Griffzungenschwerter mit ausgebauchter Zunge" kann das im Bilde 19 wiedergege-

bene in der Nähe von Bergen, Krs. Celle gefundene Stück gelten, bei dem die eingegossene Bleimasse noch recht gut erhalten und nur mit einer weißen Verwitterungsschicht von 0.5mm Stärke überzogen ist. Wie der in der Mitte der Griffzunge vorhandene Niet erweist, der 2.5–3mm aus dem Blei hervorragt, kann nur eine ebenso dicke Griffschale über dem Blei gelegen haben.

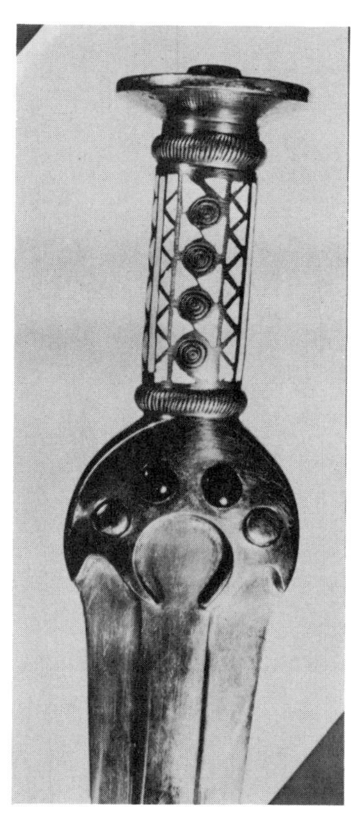

**17. Rekonstruktion des Dolches von Westerwanna. Bild 14.**

Die seitlichen Ränder der Griffzunge setzen sich an den Schultern fort und verjüngen sich vom Klingenansatz an. In der Patina ist ein halbkreisförmiger Heftabschluß mit seitlichen Rändern deutlich ausgeprägt.

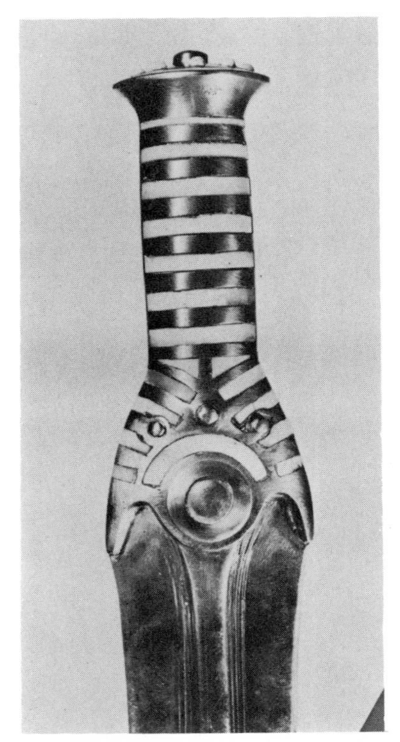

**18. Rekonstruktion des Kurzschertes Bild 15.**

Sowohl die weiße Patinaschicht des Bleies, die aus Bleikarbonat bestand, als auch die von der Heftplatte abgekratzte' grüne Patinaschicht erwiesen sich frei von Phosphatmengen, die für die Gegenwart eines Griffes aus phosphathaltigem Material sprechen könnten.

Griffzungenschwerter bei denen das zum Ausgießen verwandte Blei noch metallisch erhalten ist, sind sehr selten. Dagegen finden sich viel häufiger Stücke, die das im Bild 20 wiedergegebene Aussehen zeigen Das in Bruchstücken erhaltene 62cm lange. Schwert wurde in Wehden, Krs. Lohe gefunden und gehört dem gleichen Typ wie das zuerst erwähnte an.

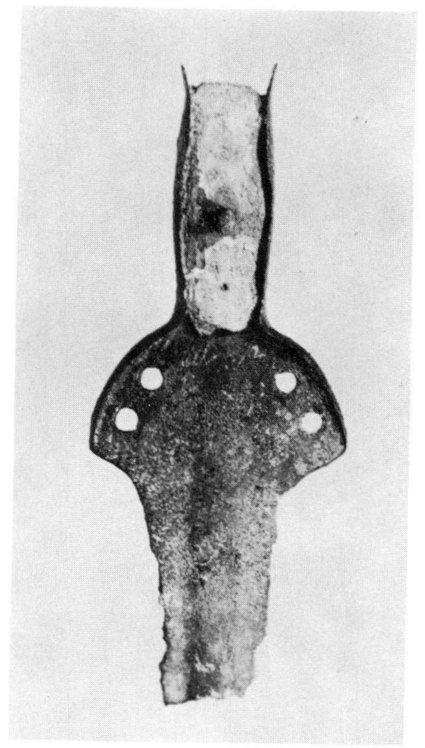

**19. Griffzungenschwert mit Bleiausguß. Fundort Bergen.**

Die Griffzunge ist mit einer bläulich weißen drusigen Masse bedeckt, in der bei 20 facher Vergrößerung hell glitzernde Kristalle hervortreten. Diese Masse liegt unmittelbar auf der Bronze, die hier oberflächlich in $Cu_2O$ übergegangen ist. Auf dem oberen Teil der Heftplatte, vor allem zwischen den Nieten, liegt eine olivgrüne, bröcklige, leicht abhebbare Patina, die oberflächlich eine ähnliche Streifenmusterstruktur aufweist, wie sie im Bild 13 gezeigt

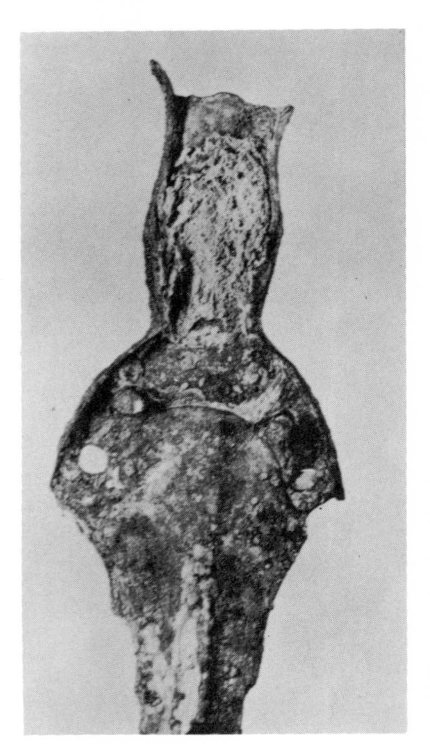

**20. Griffzungenschwert mit Bleikarbonat in der Griffzunge und Griffschalenresten auf der Heftplatte.**

ist. An den beiden unteren Nieten findet sich eine olivgrüne, höckrige Patina, die sehr fest haftet. Die seitlichen Ränder der Griffzunge tragen ebenfalls eine höckrige Patina, die auf einer roten Oxydulschicht ruht. Die Patina auf der eigentlichen Klinge ist blaugrün und haftet seht fest.

Die chemische Prüfung der aufgeführten Schichten ergibt: Die weiße, kristalline Füllung der Griffzunge ist reines Bleikarbonat mit einer Spur Silber und frei von Kupfer und Zinn. Sie führt ein wenig Sulfat, aber kein Phosphat.

Derartige Auflagen von Bleikarbonat sind bereits früher beobachtet aber in Verkennung der Tatsachen nicht als Verwitterungsprodukt von eingegossenem Blei, sondern als Bleiweiß angesprochen, das als Kittmasse für den aus Holz oder Horn bestehenden Griffbelag gedient hat. 13)

Die von der Heftplatte entnommene Patina enthält etwa 6% $P_2O_5$ und 1–3% CaO, während die von der Klinge stammende Patina kein Phosphat und nur 0.6% CaO führt.

Damit dürfte erwiesen sein, daß durch die Niete eine phosphathaltige Masse festgehalten wurde, während die Bleischicht eine dünne phosphatfreie Griffschale trug.

Bei zwei weiteren Schwertern zeigte die Patina der Heftplatte neben hohem $P_2O_5$ Gehalt (7–9%) einen CaO Gehalt von 1–2%, während in der Bleikarbonatschicht kein Phophat auftrat, sodaß für diese beiden Schwerter das Gleiche wie vorher gilt.

Von besonderem Interesse dürfte der Nachweis sein, daß im 2. Jahrt. v. Chr. bereits zur Verlegung des Schwerpunktes einer längeren Waffe zur Hand hin Blei in die Griffzunge eingegossen ist, dieses demnach bekannt gewesen sein muß. Die chemische Analyse ergibt ein relativ sehr reines Blei, das außer Spuren von Kupfer, Wismut und Antimon nahezu 0.1% an Silber führt.

c. Untersuchung einer Schmuckfibel.

Die bei Holle, Krs Marienburg aus einem Skelettgräberfeld geborgene, dem 6.–8. Jahrh. n. Chr. angehörende Fibel bietet ein Musterbeispiel dafür, wie durch eingehende verständnisvolle Untersuchungen alle Unterlagen gewonnen werden können, die für eine materialgerechte Nachbildung erforderlich sind, selbst dann, wenn die alten Metalle durch Korrosion völlig zerstört sind.

Die Ansicht der Schauseite der Fibel im Zustande der Bergung nach vorsichtiger Reinigung zeigt Bild 21.

Auf einer kreisförmigen Scheibe von 50mm Durchmesser liegen 2 rote und 3 weiße Schmucksteine von je 8mm Durchmesser. Ein roter Stein und der Mittelstein sind verloren gegangen, von letzterem ist nur die Fassung vorhanden, die mit einer weißlichen Masse gefüllt ist. Den Rand des Stückes bildet ein hochgetriebener Perlenrand, mit dem auch jede der Steinfassungen umschlossen ist, während um die zentrale Fassung ein doppelter Kranz liegt. An einzelnen besser erhaltenen Stellen läßt sich erkennen, daß parallel dem äußeren Perlrand ein Ring von 5mm Breite hochgetrieben und mit zahlreichen Querrippen versehen ist. Die zwischen den Perlringen der Fassungen liegenden dreieckigen Metallflächen wurden zu stumpfen Pyramiden erhöht.

Die ursprüngliche Metallplatte besteht heute nur aus einer blaugrünen, dünnen Patinaschicht, Kupferkarbonat mit 0.5% PbO und 0.6% $SnO_2$, führt reichlich Eisen (6.4% $Fe_2O_3$) und mehr als 20% Bodenbestandteile.

Unter der Patinaschicht der Schauplatte liegt eine feste, dünne Eisenoxydschicht, die rd. 15% CuO neben 64% $Fe_2O_3$ enthält. Blei, Zinn und Bodenbestandteile sind nur spurenweise vorhanden.* Sie besteht demnach aus Eisen (III) oxydhydrat, das durch Beimischung von etwas Patina aus der überliegenden Schicht verunreinigt ist.

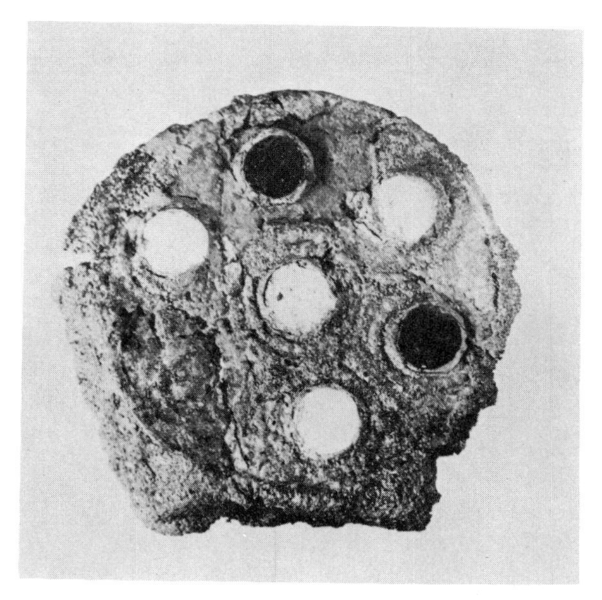

**21. Scheibenfibel von Holle, Schauseite, Originalzustand, etwa nat. Größe.**

An Stellen, wo die ursprüngliche Metallplatte hochgetrieben war und unter den "Steinen" läßt sich zwischen der grünen Patina und der rotbraunen Eisenoxydschicht eine scharf begrenzte bläulich weiße Zwischenlange beobachten, die nach mikrochemischem Befund aus weißem Ton und Calciumkarbonat besteht, die durch eingewandertes Kupferkarbonat (etwa 5%) verfärbt ist und außerdem noch Bleikarbonat (etwa 2%) enthält. Für die

Anwesenheit des Tones spricht der Titangehalt, der dem aus $Al_2O_3$ und $SiO_2$ errechnetem Tongehalt entspricht.

Die roten "Schmucksteine" sind 0.7–0.75mm dicke, rund zugeschliffene durch Kupfer rot gefärbte Gläser.

**22. Scheibenfibel von Holle, Rückseite, Originalzustand. Mit Glasperlen und Gewebezeichnung.**

Die weißen "Schmucksteine" sind 0.85–0.9mm stark und bestehen aus Calciumkarbonat mit geringer Beimischung von Phosphat. Im Querschnitt zeigen sie bei mikroskopischer Betrachtung einen charakteristischen Aufbau, der auf die Verwendung von "Perlmutter" hinweist. Diese Vermutung wird durch röntgenographische Aufnahme bestätigt, bei der ein Debye-Sherrer-Diagramm erhalten wurde, das sich völlig mit dem des Aragonits deckt. Da aber Perlmutter ebenfalls das Debyeogramm des Aragonits zeigt 15), und das mikroskopische Bild des

Querschnittes gleichfalls dafür spricht, dürfte das Vorliegen von "Perlmutter" gesichert sein.

Die Rückseite der Fibel zeigt Bild 22. Am oberen Rande erkennt man 2 stabförmige und 2 runde Glasperlen, die erweisen, daß das Schmuckstück nicht als Fibel sondern als Anhänger an einer Halskette getragen ist, was auch seiner Lage auf dem Skelett entsprechen dürfte.

**23. "Gewebereste" auf der Rückseite der Fibel. Seitliche Beleuchtung; Vergr. 12 x.**

Vor allem ist aber die Oberflächenstruktur der Masse auffallend. Sie zeigt sehr deutlich den Abdruck eines Gewebes in Leinenbindung, was noch deutlicher auf dem bei schwacher Vergrößerung und schräger Beleuchtung aufgenommenem Bilde 23 zu erkennen ist.

Die chemische Prüfung der braunen Masse zeigt, daß sie im wesentlichen aus Eisenoxydhydrat besteht, nur wenig Bodenbestandteile (etwa 4%), rd. 1% Kupferoxyd aber kein Blei enthält.

Damit sind alle Unterlagen vorhanden, die für die materialgerechte Rekonstruktion des Fundstückes erforderlich ist.

1. Die Vorderseite des Schmuckstückes muß ein dünnes Kupferblech gewesen sein. Ein Bronzeblech scheidet aus, da der Zinndioxydgehalt der Patina, in der sie ja dieses anreichern mußte, viel zu gering ist. Ebenso ist ein Messingblech auszuschließen, da bei allen Analysen keine Spur Zink gefunden wurde.

2. Drei der Schmucksteine sind rot gefärbte Gläser, drei weitere zugeschliffene Perlmutterscheibchen. Über den Mittelstein ist nichts zu sagen.

3. Die Erhöhungen der getriebenen Kupferplatte und die eingesetzten Steine sind mit einer aus weißem Ton und Calciumkarbonat gemischten Füllmasse unterlegt, die wahrscheinlich durch Beimischung eines organischen Bindemittels (Wachs oder Harz) verfestigt wurde.

4. Die Bodenplatte bestand aus einem runden Eisenblech. Die aus diesem bei der Verwitterung gebildeten Eisen (II) salzlösungen durchtränkten das unterliegende Gewebe. Bei der später einsetzenden Oxydation und Hydrolyse wurde in und auf den Fasern Eisen (III) hydroxyd abgeschieden das verhärtete und so die Struktur des Gewebes in allen Einzelheiten erhielt. Irgendwelche Reste der ursprünglichen Faser sind nicht mehr vorhanden. Es liegt eine Pseudomorphose von Eisenhydroxyd nach einem Leinengewebe vor.

Nach den Untersuchungsbefunden kann die materialgerechte Nachbildung des Schmuckstückes in folgender Weise durchgeführt werden.

Aus einer dünnen Kupferplatte wird nach dem noch erkennbaren Muster die Schmuckseite getrieben, wobei zunächst die Perlenreihen mit einer passenden Punze von der Rückseite her eingeschlagen werden. Dann werden die Fassungen für die "Steine" hoch getrieben und ihr oberer Rand etwas nach innen umgebörtelt, sodaß die bereits zugeschliffenen Glas-und Perlmutterplättchen festgehalten werden. Nach dem einpassen der Scheibchen werden die Erhöhungen der Kupferplatte und die Hohlräume unter den Steinen mit der nach Zusatz von Wachs oder Harz und Erwärmen plastisch gemachten Füllmasse hinterlegt. Die so vorbereitete Schmuckplatte wird mit einer, einige mm kleineren Eisenplatte hinterlegt, die auch die eiserne Ose zum Befestigen der Kette trägt, und der überstehende Rand der Schmuckplatte über die Eisenplatte gebörtelt.

**24. Materialgerechte Rekonstruktion der Fibel. Etwa nat. Größe.**

Einen infolge der fehlenden Farben nur unvollständigen Eindruck der Rekonstruktion vermittelt das Bild 24. Der rötliche Ton des glänzend polierten Kupfers, das Farbenspiel der Perlmutterplättchen und die tiefroten Glasscheibchen vereinigen sich zu einem harmonischen Farbenbild von einer künstlerischen Wirkung, von der das Original im jetzigen Zustande nichts ahnen läßt.

## ZUSAMMENFASSUNG

Es werden die Ergebnisse der chemischen Untersuchung einer Reihe von Patinaproben mitgeteilt, die von einigen der in Nordwest-Deutschland gefundenen vorgeschichtlichen Bronzen stammen. Aus den Analysenbefunden, die durch experimentelle Versuche ergänzt sind, wird die Möglichkeit der Verwitterung von Bronzen bis zu einer reinen Zinnsäure gezeigt.

Beschreibung und Bilder derartiger "Pseudomorphosen von Zinnsäure nach Bronzen" sowie ihre chemische Zusammensetzung werden ebenfalls gebracht.

Die Untersuchung der von Bronzedolchen -und Schwertern entnommenen Patina auf den Phosphatgehalt gestattet Aussagen über die Art der für die heute vergangenen Griffe in der Zeit zwischen 1500 und 1200 v. Chr. verwandten Materialien, wobei, vielfach Knochen bezw. Hirschhorn in Betracht kamen.

Durch Nebeneinanderstellen der Bilder von Fundstücken und material gerechten Nachbildungen wird die ursprüngliche Schönheit einiger alter Waffen gezeigt, die nach dem Aussehen der Fundstücke selbst kaum zu erwarten war.

Aus dem Auftreten von Bleikarbonat in den Griffzungen einiger Schwerter wird auf die Verwendung von Blei zur Verlagerung des Schwerpunktes geschlossen, was durch ein Griffstück mit noch vorhandenem Bleieinguß bestätigt wird. Benutzt ist ein relativ reines Blei mit geringem Silbergehalt, dessen Darstellung und Verwendung demnach für die angegebene Zeitspanne erwiesen ist.

Zum Schluß wird über die Untersuchung eines völlig oxydierten Schmuckstückes berichtet, nach deren Ergebnissen eine materialgerechte Rekonstruktion möglich war, die im Lichtbild wiedergegeben ist.

Die vorstehenden Mitteilungen sollen zeigen, daß die Auswertung der Ergebnisse naturwissenschaftlicher Untersuchungen, die bei Verwendung neuzeitlicher Arbeitsverfahren auch mit geringsten Materialmengen durchführbar sind, dem Kulturhistoriker wertvolle Erkenntnisse vermitteln kann, die ohne diese nicht zu gewinnen sind.

## LITERATURZITATE

1. W. Geilmann u. K. Meisel, Libethenit, ein Mineral der Patinabildung Nachrichtenblatt für Deutsche Vorzeit *18* 208–212 (1942)

2. W. Geilmann, Verwitterung von Bronzen in Sandboden Angewandte Chemie *58* 201–211 (1956)

3. W. D. Asmus, Germania *23* 168 (1939)

4. O. v. Olshausen, Verh. Berliner Ges. Antropologie, Etnographie u. Urgeschichte (1883). 88 und 467 (1884), 524 und (1897) 344 und 352.

5. G. A. Rosenberg, Antiquités en fer et en bronze, leur transformation dans la terre contenant de l'acide carbonique et des chlorures et leur conversation. Kopenhagen (1917)

6. W. Geilmann u. Waltraud Gebauhr, Über einige Leinengewebe aus Bronze-gefäßen des 3. Jahrh. n. Chr. aus Niedersachsen. Die Kunde Neue Folge 10. Seite 260–269 (1959)

7. J. Schuler, Dingl-Polytechn. J. 232-333 (1879)

8. A. Kröhnke, Dissertat. Kiel 1892

9. E. v. Bibra, Bronzen und Kupferlegierungen der alten und älteren Völker, Erlangen (1869). L. R. von Fellenberg, Mitt. Naturforschend. Gesellsch. Bern (1865) 12

10. E. E. Free, The Solubility of precipitated Basic Copper-Carbonate in Solutions of Carbon Dioxide. J. Amer. chem. Soc. *30* 1366 (1908).

11. W. Mecklenburg, Zur Isometrie der Zinnsäure Z. anorg. Chem. *74* (1912) 215

12. W. Geilmann, Chemische Untersuchungen an vorgeschichtlichen Bronzwaffen Niedersachsens. Mit prähistorischen Erläuterungen von K. H. Jacob-Friesen.-Nachrichten der Gesellschaft der Wissensch. zu Göttingen, Philologisch-Historische Klasse, Fachgruppe I Neue Folge II Nr. 3 (1937)-47-66

13. Sprockhoff, Römisch-germanische Forschungen Bd. 5 Verlag Walter de Gruyter, Berlin 1931. Zeitschr. f. Ethnogie *15* (1883) S. 105.

14. W. Geilmann, Untersuchung der Scheibenfibel von Holle Die Kunde, Jahrg. 6 4–13 (1938)

15. W. Fr. Eppler, W. M. Lehmann und H. Rose, Deutsche Goldschmiedezeitung Nr. 16 (1925)

# New Methods in the Investigation of Ancient Mortars

By Hanna Jedrzejewska
National Museum, Warsaw

## INTRODUCTION

The examination of ancient mortars is carried out mostly from the technological point of view. The number of specimens examined is here relatively small, limited by the needs as well as by the expensive and time consuming methods of analysis. No true comparative studies have been done on mortars, in fact, no useful correlations were probably ever expected here.[1] Besides, the number of specimens to be examined would have to run into thousands, and no one could afford the expense of so many analyses. The problem of collecting the necessary amounts of authentic samples seemed also discouraging.

This last difficulty was solved during the war. The great number of ruined ancient buildings became an excellent source of interesting samples, well suited for a systematic comparative investigation. There was also a considerable supply of samples from archeological relics of ancient architecture. The investigation of Polish mortars was started by the writer in 1950. At first, the basic methodic principles had to be worked out. Also a simplified analytical procedure was needed for the directly relevant properties of mortars. This had to be quick and inexpensive.

As a whole, the new method seems to be serving its purpose well. Over 1500 mortars have already been examined, and much interesting material discovered, obviously pointing out the fact that our knowledge on the subject is far from complete.

## THE GENERAL PRINCIPLES

The main purpose of the research was to find the eventual correlations among the kinds of mortar, the time and place of their use, and the particular groups of builders using those mortars. Any other interesting information gathered in the research had of course also to be included in the final results.

There is no need, in research of this kind, to determine all the possible properties of the sample, as is done in technological examination, but only those properties that are significant for the intended purpose. It is only a problem to decide which of them are significant.

Unfortunately, the properties of mortars usually determined in the conventional analyses [2] were not suitable here, and it became necessary to look for properties better filling the requirements of comparative research. To have a starting point for these considerations it was assumed as a working hypothesis, that, according to medieval traditions, every group of builders had its own individual recipes for mortar, and that certain differences might be expected in these particular recipes.

Generally, a recipe is defined by the kinds of ingredients, their amounts, and the method used to prepare the mixture. Thus it is in these three particular points that differences in recipes for mortars could have existed. But in the process of hardening, chemical reactions are known to take place in the mortar. Now, if some of the original differences in the recipe could still be traced in the hardened mortar, this could give a useful key for comparative studies.

## THE MORTAR

As is well known, old mortars belong mostly to the class of hydraulic lime mortars. They have lime as a basic constituent, but also a certain proportion of hydraulic components,[3] and an inert filler (mostly sand). During the hardening process of such a mortar, part of the lime becomes bound to the active oxides (the hydraulic components) to form various complex silicates. The remaining part of lime slowly combines with carbon dioxide to form calcium carbonate. Sand remains practically unchanged.[4] The proportions of these three constituents in the hardened mortar will depend on the kinds of ingredients originally used, their proportions and the method of preparing the mortar. In other words, the amounts of these three constituents will depend directly on the original recipe. Therefore, it was decided to determine their contents in the investigated samples, and to use the corresponding values as a direct means of preliminary classification. As it happens, the method is as well suited for other kinds of mortar, for example calcium carbonate mortars,[5] gypsum mortars, clay, etc.

## THE ANALYSIS

No very sensitive methods were required here. A simplified semi-quantitative procedure proved accurate enough for the analysed materials, often markedly lacking in

Fig. 1. Apparatus for the volumetric determination of carbon dioxide in mortars. (1) Crucible with sample. (2) Cork. (3) Mercury. (4) Rubber stopper. (5) Dropper pipette with concentrated HC1. (6) Gas burette (100 cc.).

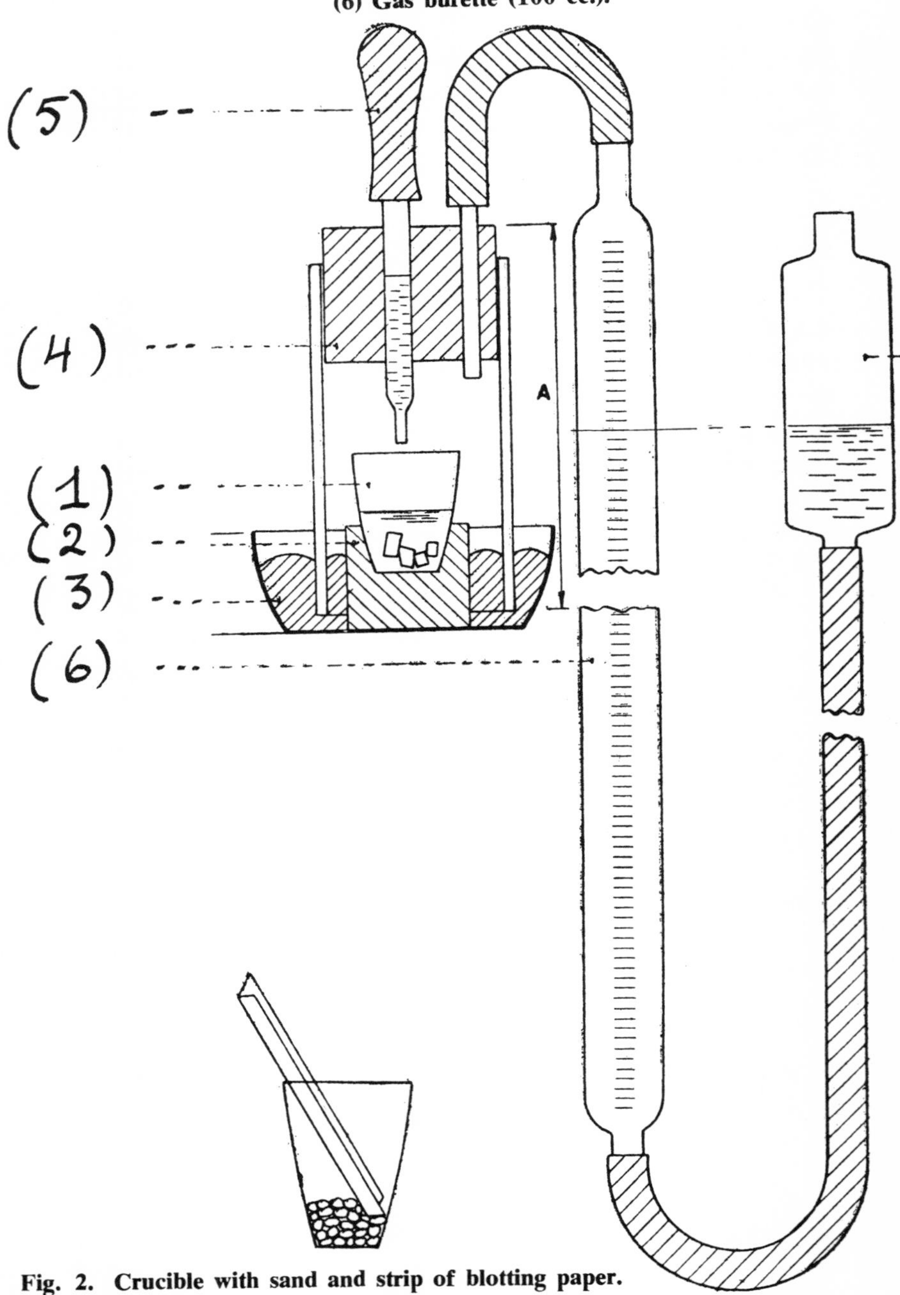


Fig. 2. Crucible with sand and strip of blotting paper.

uniformity. The relative proportions of calcium carbonate, sand and complex silicates (solubles) were determined in one single operation of dissolving the sample in hydrochloric acid in a special apparatus (see Experimental), and expressed in w/w percentages of the whole sample. These values were directly used for classification. This seemingly over-simplified treatment proved to be very convenient and useful in the comparative investigations. It was also simple enough to be used for large numbers of specimens.

The three basic values from the analysis were very different for various kinds of mortars. The proportion of calcium carbonate varied, approximately from 0 to 95 percent, of solubles from 2 to 35 percent, of sand from 0 to 90 percent. This gave ample margin for deviations caused by non-uniform structures of certain mortars, and also allowed a good differentiation among various groups of mortars (see Results).

If the agreement in results was good, only two determinations were made for one sample. In cases of greater deviation, two more determinations were made. No average contents were calculated because the differences in the particular analyses were a useful indication of the uniformity of the mortar. The reproducibility of results was good, even after several years.

Also, the kind of sand, the color of the solution in HC1, the kind of suspensions, and the results of visual examination of the sample were noted as a secondary help to the examination.

## SAMPLING PRINCIPLES

The elementary principle of sampling is that the sample has to be truly representative of what is being analyzed. If the samples are not just in that respect, even the best method of analysis is valueless, and the interpretation of data leads to errors.

Therefore, great care must be given to the problem of sampling mortars. First of all, the mortar must be authentic, so that it is best taken by someone well acquainted with the building. It must also be well preserved, not powdered, and taken from the interior of the wall to avoid changes due to weathering, infiltration, or repair. Its quantity may vary from 5 to 25 g. to have sufficient material for analyses, for the collection, and eventually for more detailed examination. The exact spot where the sample was taken must be noted carefully.

But besides all these precautions with every particular sample, it is also necessary to prove that the composition of the sample is truly representative of the mortar in the investigated element of architecture. This is done by taking several samples from various parts of the investigated element. If the composition of all these samples remains reasonably constant, it may be considered as representative of the whole element. But it must always be remembered that single samples can not be used as characteristic of anything.

To make the sampling problems complete, that is, to connect the sample with architecture or history, it is also necessary to have the building itself well analyzed by specialists, to decide as accurately as possible on dating, influences, style, connections, type of building, purpose, etc.

## RESULTS AND CONCLUSIONS

The method gives only a very general picture of a mortar, and, therefore, the interpretation of results must be very careful and cautious. It is also too early to draw any over-general conclusions about definite correlations or chronological patterns. Nevertheless, there are, by now, several facts and possibilities, that may be of interest to investigators in the fields of technology, of archaelogy and of ancient architecture. Some of these facts will be briefly noted here and commented upon.

*The classes of mortars.* The analytical procedure used here to investigate mortars gives no detailed information on the kinds of components except what can be detected by visual observation, but it makes possible a classification of mortars into groups of definite characteristic properties connected with the original recipe (Table 1):

*Gypsum mortars* are easily detected by the absence of sand (in most cases) and a very low content of carbonates. Sulphates have to be confirmed by chemical tests. Gypsum mortars usually are slightly porous, well preserved, very hard, of colours from grey to rose, occasionally almost white. They were found in Poland in certain relics of architecture of the eleventh and twelfth centuries, and seem to be connected with a definite group of builders. But nothing more is at present known about them. Similar gypsum mortars were used approximately at the same period in South Germany and in France, but not exclusively. Gypsum mortars seem to be used exclusively in Ancient Egypt until the Ptolemaic period.[6] It would be interesting to have more examples of the use of this unusual mortar and to trace its history.

*Carbonate mortars* are lime mortars with a calcium carbonate filler. They can easily be detected by the absence (or very low content) of sand, and very high proportion of carbonates. They are usually very hard, microcrystalline, slightly porous. They have been found in Poland in certain buildings and relics, most probably of the eleventh century. Nothing definite is known at present about who used this kind of mortar in Poland, but, again, there are indications of some particular group of builders. In general, not much is known about this kind of mortar. Gettens [7] called attention to lime mortars in the walls of Aegean relics of pure calcium carbonate with no detectable filler such as limestone, marble, etc., but very hard and compact, which is an effect impossible to get with pure lime. Taylor [8] mentions that the Greeks used lime with only a little sand (calcium carbonate filler?). The writer has found examples of carbonate mortars in the Greek relics of Mirmeki in Crimea.[9] Jun [10] discovered carbonate mortars on South Russian territory (twelfth century), and he discusses in detail their excellent properties. It would perhaps be interesting to trace more of the origin and history of carbonate mortars.

*Lime mortars* are mortars made of lime and sand, as chief constituents. They may have some additional ingredients to improve their qualities (see Note 3). Exceptionally, crushed ceramic material is added instead of sand.

This group is very large and has to be further divided into sub-groups. In Poland the most popular are the four groups of lime mortars presented in Table 1.

The first two groups, with a low content of sand and a high proportion of calcium carbonate, but one of them with a very high amount of solubles, are certainly very different in their technological properties, and were obviously used at different times, and by other groups of builders (eleventh, thirteenth, sixteenth century, as dated by the architecture). One of them seems to be closely accompanying the Cystersians' buildings.

The third sub-group of lime mortars in Poland was very popular in the Romanesque architecture of the twelfth century. The proportions of ingredients are here rather closely maintained, but the kinds of ingredients and the appearance of the mortar may vary. In the walls of the same time and origin the mortar is as a rule of one kind only.

The fourth sub-group of lime mortars is characterized by a very high content of sand. They are mostly dark grey in colour. Only one sample was bright rose. They were very popular in the Gothic architecture of the fourteenth and fifteenth century. Some of them are very hard and well preserved, others have marked tendencies to disintegrate. The technological differences are marked between these mortars, again, the kind of mortar used for one particular building seems to be well standardized.

According to the working hypothesis (see General Principles), the regularities in mortar composition should become complicated after the sixteenth century owing to the gradual decline of the medieval traditions. This was very strongly confirmed in the investigation. In fact, no comparative studies of mortars of later date were possible.

*The Ingredients and Recipes.* Detailed chemical and petrographic examination is necessary here, of course, as the comparative analysis only gives a preliminary indication of possible components. But the material already collected from literature, observation and experiments is perhaps worth a brief review.

The ingredients found in Polish mortars by experiment or by visual examination were: (a) binder—lime or gypsum (clay mortars were not investigated); (b) hydraulic addition —clay, crushed brick, powdered stone material, crushed iron ore; (c) filler—sand, crushed lime-stone, crushed fragments of gypsum mortar, crushed brick. A detailed petrographic examination of selected specimens will of course reveal more of the original ingredients, especially of hydraulic properties. The supposed addition of organic materials to mortars, as sometimes mentioned in literature, was not confirmed beyond doubt.

But it seems that far more important than the ingredients are the methods used to prepare the mixture and, especially, the lime.

A considerable amount of heat is evolved in the reaction between calcium oxide (quick-lime) and water. According to Lange [11], the temperature may rise to 270–300°C, and even over 450°C in special conditions. This heat is practically lost when quicklime is slaked in open boxes. But if it were slaked in a closed space, the amount of heat evolved and the overheated steam, under pressure at that, would exert a very strong chemical action on any siliceous materials present, bringing them, at least partly, into solution. Lime, slaked by this method, would have quite different properties than lime from the same raw materials slaked by the open-box method. This possibility is very strongly reflected in a series of German patents of the nineteenth century at the early period of Portland cement.[12] Prescriptions are given for various lime mortars and cements prepared from quicklime slaked in closed containers in the presence of different siliceous materials such as powdered low-burnt brick, special kinds of sand, ashes of brown coal, powdered pumice stone, trass, iron slag, diatomaceous earth, clay, Portland cement, asbestos, etc. All this led to mortars of greatly improved qualities due to the presence of active hydraulic components.

There are certain indications that such a method could have been used for the preparation of ancient mortars, much earlier than nineteenth century. Berger [13] gives a description of a method of slaking quicklime in heaps well covered with sand, and using only small amounts of water, to be found in *Kunst und Weckschul* (1707), p. 723. According to Berger the method is probably based on much older traditions. The siliceous materials could be present in quicklime naturally from clay and other impurities, or they could be added intentionally before slaking. This could perhaps be another cause for some extraordinary properties of ancient mortars, so mysteriously impossible to reproduce, even when, apparently, the same ingredients were used. Experiments carried out by the writer, with lime slaked in the presence of common clay, resulted in a mortar very similar to the kind of old mortars heretofore impossible to reproduce.

Another method used in ancient times to improve the properties of lime mortars was to prepare a mixture of well-slaked lime, sand and, eventually, hydraulic ingredients, and to add to this, just before use, some proportions of freshly burnt quicklime. Such a "hot lime" had to be used immediately. According to Lange,[14] this method was used by the Romans. It was certainly known and used in later times.

In the interesting collection of prescriptions in Lange's book, recipes for gypsum mortars are also included, concerning both the estrich kind, made of gypsum—burnt in much higher temperatures than plaster of Paris—and the low-burnt kind with various additions to make it harder. It seems that raw gypsum stone was added to these mortars as filler. In such a case the presence of small amounts of calcium carbonate would not directly indicate estrich gypsum as it might as well come from impurity in the stone.

Concerning the carbonate mortars, there is only one interesting German patent (D.R.P. 303 319)[15] for a mortar of hydraulic properties prepared from pure, old, well-carbonized lime slaked together with fresh quicklime. Perhaps this could be connected with the lime of mysterious properties from the Aegean walls. The problem of the excellent properties of carbonate mortars was investigated in more detail by Jung.[16] He supposed that the crystals of calcium carbonate present in the filler may act as nuclei for the crystallization of carbonized lime.

Between carbonate mortars and lime mortars with siliceous ingredients, there is a special group of lime mortars with some amounts of sand filler, but also with an unusually high content of carbonates which suggests the presence of carbonates in the freshly prepared mixture. These carbonates might have been added intentionally, which was probably often done, but they could have come also from incompletely burnt limestone.

Besides the problems of various special ingredients and of different methods of using the quicklime, also the amounts of water in the ready mixture could be different than those used at present. Vitruvius gives a short description of how the Greeks used to mix and beat their mortar which was well known for its hardness and excellent properties. Also, as observed now, the carbonate and gypsum mortars have a porous and sometimes stratified structure which concords with the suggestion of the original very stiff consistency of certain mortars.

All this should be taken into consideration when attempts to reproduce old mortars are made. Some of these recipes could also be very useful for certain problems in conservation, where there is need of materials with better properties than simple lime mortars, and where Portland cement cannot be used.

*Archaeology, architecture, history.* With careful attention to methodical principles, the different kinds of mortars can be correlated with the architecture (see Sampling

Principles) of various periods and origin. By comparison of mortars it is possible to detect, for example, which elements of one or more buildings may belong to the same period. If some of these elements are dated, it will date the others too. But one must be very careful with dating, and support it with investigations in other fields. The composition of a mortar alone cannot be, of course, used for dating.

It is also possible to study only one particular kind of mortar, and to follow the geographical and chronological pattern of its use. It may be tried, too, to connect the various kinds of mortars with various groups of builders, beginning with objects with a well known history. The chronology of mortars in various regions may also be well worth investigating.

Some of the investigations carried out by the writer in this latter connection may perhaps be well illustrated by the example of Wislica, Poland. This is a small town, now unimportant, but with very ancient traditions. Extensive archaeological excavations carried out for several years around the Gothic Collegiate and under its floor, have resulted in the discovery of several relics of earlier architecture: in the yard, it was a small early church (tenth century?), with a later addition (twelfth century?). Under the Collegiate, there were remains of an early Romanesque crypt with a very finely decorated floor, and walls of a later, big Romanesque church with two towers. During the excavations samples of mortars from all discovered fragments were systematically taken and analyzed according to the basic principles of the comparative method. Very marked differences were found between mortars from various periods, but with very well defined types of mortars for every particular object.

Some analyses are presented in Table 2, as example. In

general, it was found that the mortar from the earliest church was not typical for mortars of the probable dating (it was dated on the base of archaeological finds). But the same kind of mortar was found in certain buildings of the same date in other regions of Poland, and supposed to be built by the Benedictines. The addition to this church and the Crypt were built with gypsum mortar of two kinds, grey and rose. The decorated floor-plate was of gypsum. Some other objects in the near vicinity, probably of similar date, were also made with gypsum. The same mortar was found in three objects in other regions of Poland, of a probably somewhat earlier date (eleventh century?). The recipe might have come from South Germany or France, but nothing more definite is known at present about who did erect all the buildings.

The big Romanesque church is under study. Preliminary samples may point to a certain difference in period between the main building and the towers. The mortar of the towers bears a very marked resemblance to mortars of a Cystersian church in the near vicinity. The mortars of the Gothic Collegiate and of the adjoining Gothic building are rather typical for mortars commonly used at that time in all the territory of Central Poland.

The research in Wislica is being carried on systematically by various specialists, and perhaps some problems will find their solution in further historic and architectural investigations. But one thing is already obvious: that quite different groups of builders were active here at various periods. There are other examples in Poland of architecture where the builders did not change (the monastery was in the same hands for hundreds of years), and where the kind of mortar remained practically the same.

*Plasters and grounds for murals.* The general rules for

mortars are only partly relevant to plasters and grounds of mural paintings, as in most cases the grounds were specially prepared for painting purposes. The proportions as well as the ingredients could be quite different here from the mortar in the walls. But here the comparative method is good for a preliminary analysis, or to differentiate paintings of various periods. The abundant amount of soluble silicates in certain kinds of mortars, when these are used as fresco grounds, may act as binders for pigments. Perhaps the mysterious cases of murals that are not frescoes, and at the same time, where the organic medium cannot be determined, are just examples of this.

## EXPERIMENTAL

The comparative study of ancient mortars will never be successfully carried out unless the same method of examination is used by all authors. The standardized procedure proposed by the writer was first published in *Studies in Conservation.*[17] It is quoted here and the figures supplied.

The determination of the three basic values of a mortar is performed by dissolving the sample in hydrochloric acid (HCl) in a special apparatus which measures the volume of carbon dioxide ($CO_2$) (Fig.1). A well chosen fragment of mortar of about 0.4 to 0.8 gm. is first crushed, then placed in a small porcelain crucible, dried, weighed, covered with about 3 ml. of water, and placed in position (2). The dropper pipette (5) is filled with about 3 ml. of conc. HCl from a test tube. The apparatus is closed by putting the upper part (*A*) into the mercury (3) and fixing with a clamp. The water levels in the two arms of the burette are equalized, and the volume recorded. HCl is gradually dropped from the pipette, and $CO_2$ is evolved. When the reaction is complete, the new volume in the burette is recorded, the crucible is taken out, and the solution, after stirring with a glass rod, is immediately transferred to a dry test tube, together with light suspensions and precipitates. (No water should be added.) The sand is then washed with water several times (the water discarded) and left for twenty-four hours in the crucible to become air-dry. No heat is used. A strip of blotting paper put into the crucible helps the drying (Fig.2). The dry sand is weighed.

The amount of carbonates (calculated as calcium carbonate), corresponding to the amount of $CO_2$ evolved, is recorded directly as w/w percentage of the whole sample. The amount of sand is expressed in the same manner. The solubles are calculated by summing up the percentages of carbonates and sand, and subtracting them from 100 per cent.[18]

*Notes:* (1) The powdered samples are left in crucibles for twenty-four hours to become air-dry, before the analysis is done. Gentle infra-red heating may be used to dry both the samples and the sand.

(2) A small error is incurred by not taking into consideration any magnesium carbonate that may be present.

(3) Additional observations, including visual and microscopic examination of original mortar, sand and precipitates, may help in the classification of closely similar mortars.

(4) The analyses are made in sets of twenty, with an average time of thirty minutes for one analysis. At least two analyses of each sample should be taken.

The experimental error in the sand determination is from 1–3 percent, calculated on the average sand content in the mortar. Allowing for this, variations in the sand content afford an identification of the homogeneity of the mortar. It is sometimes useful to consider this feature in more general studies. The experimental error in the volumetric determination of carbonates amounts to about ± 0.5 ml. of carbon dioxide.

To save time, rather small samples were analyzed. For mortars of a uniform structure the results were satisfactory even for samples below 0.1 gm. For coarse mortars the size of analyzed fragments should not be less than 0.2 gm. Usually about 0.4 to 0.8 gm. samples were taken.

## NOTES

1. F. S. Taylor. *A History of Industrial Chemistry* (London, 1957), pp. 55–58, a short survey of various kinds of mortars used in ancient times which ends with the following statement: "We may say, then, that the typical cement and mortar from the period before the 18th century was the familiar mixture of lime, sand and water."

2. The chemical composition of a mortar is as a rule expressed in proportions of oxides (CaO, MgO, $Al_2O_3$, $Fe_2O_3$, $SiO_2$, etc). Carbon dioxide is usually determined by calcining the sample at 900°C., seldom by the volumetric methods. For physical structure, mortars are examined petrographically as thin polished sections. The kinds and grades of sand are also studied. Mechanical and physical properties are sometimes included. More advanced investigations are aided by spectrographic methods, thermography, and electron microscopy. The methods are far from standardized, so that general comparative studies are often impossible. Besides, all the properties of mortars, basic, secondary, and even accidental, are examined together, without considering their real technological importance. For details see:

W. W. Scott. *Standard methods of chemical analysis,* Vol. II (London, 1956), pp. 1597–1619. A. Lucas. *Ancient Egyptian Materials and Industries* (London, 1948). E. Berger. *Beiträge zur Entwicklungsgeschichte der Maltechnik* (Munich, 1904). P. Duel and R. J. Gettens, "A Review of the Problem of Aegean Wall Painting," *Technical Studies in the Field of Fine Arts,* 10, No. 4 (1942), pp. 179–223. F. Müller-Skjöld. "Über antike Wandputze." *Angewandte Chemie,* 33 (1940), pp. 139–141. W. N. Jung. *The Technology of Mortars* (Osnowy technologii wiazuczych wieszczestw) (Moscow, 1953). I. L. Znaczko-Jaworski, "Experimental Research on Ancient Mortars and Binding Materials," *The Quarterly of History of Science and Technology* (Warsaw), 1958, No. 3, pp. 377–407. K. Biehl. "Beiträge zur Kenntnis alter Römermörtel," *Tonindustrie Zeitung,* No. 10, 1927, pp. 139–143; No. 9, 1928, pp. 346–8; No. 22, 1929, pp. 449–57.

The above publications are no more than a small selection on the subject of mortars but chosen to give a picture of the variety of methods which have been used to study this problem.

3. Hydraulic components contain active silica and oxides of iron and aluminum, all capable of reacting with calcium oxide (hydroxide) to form various complex silicates and aluminosilicates. Their composition depends both on the components and the preparation of the mortar. The most common hydraulic additions are: certain volcanic earths, powdered brick, clay, diatomaceous earth. Water causes slight hydrolysis of these silicates, resulting in an alkaline reaction, sometimes mistaken for the presence of free calcium hydroxide. Jung (see Note 2) discusses the problem in detail. Working with alcoholic phenolphthalein, he did not detect the presence of free lime in any of the examined mortars. The same result was obtained for the old Polish mortars.

4. There seems to be a controversy concerning the possible reaction between silica in sand, and calcium hydroxide in slaked lime. Znaczko-Jaworski (see Note 2), using petrographic analysis of quartz sand grains in ancient mortars, found their surface strongly corroded. He supposes this to be the effect of the corrosive action of calcium hydroxide over a very long time. But, more probably, it could be caused by the strong action of overheated steam and calcium oxide during slaking in covered heaps.

5. Calcium carbonate mortars are lime mortars with a calcium carbonate filler (e.g. limestone, marble, old carbonized lime, shells).

6. A. Lucas, *op. cit.*

7. P. Duel and R. J. Gettens, *op. cit.*

8. Taylor, *op. cit.*

9. H. Jedrzejewska, "Preliminary Examination of Mortars from Mirmecki." Contribution to *Kazimierz Michalowski, Mirmecki* (Wroclaw, 1958.

10. W. N. Jung, *op. cit.*

11. Otto Lange, *Chemisch-Technische Vorschriften* (Leipzig, 1923), pp. 817–853.

12. *Ibid.*

13. Taylor, *op. cit.*

14. Lange, *op. cit.*

15. *Ibid.*

16. *Ibid.*

17. H. Jedrzejewska, "Old Mortars in Poland: A New Method of Investigation," *Studies in Conservation,* 1960 Vol. 5, No. 4.

18. Sometimes there are other causes for the obtained results of analysis. The amount of carbon dioxide evolved, for example, depends not only on amounts of carbonized lime, but also on the presence of a carbonate filler. The amount of solubles is influenced not only by the acid–soluble silicates, but also by other ingredients soluble in hydrochloric acid—those which dissolve with evolution of gas (e.g. gypsum or uncarbonized lime), and those whose dissolution leads to a light precipitate (as with certain organic substances).

TABLE 1.

Classes of Mortars in Poland

| *Class of mortar* | | *% Sand* | *% $CaCo_3$* | *% Sol.* | *Approx. Chronology* |
|---|---|---|---|---|---|
| *Gypsum* | | 0 | 1–15 | (partly) | 11–12th c. |
| *Carbonate* | | 0 | 90–98 | 2–10 | 11th c. |
| *Lime and Sand:* | 1 | 30–40 | 30–40 | 20–35 | |
| | 2 | 30–50 | 30–50 | 10–20 | 11, 13, 16th c. |
| | 3 | 60–65 | 20–30 | 10–15 | 12th c. |
| | 4 | 70–90 | 8–20 | 3–10 | 14–15th c. |
| *Lime and Sand* (1 : 3) | | ca 80 | ca 15 | ca 5* | 20th c. |

* The contemporary lime mortars have a very low content of acid-soluble iron compounds and silicates (unless mixed with portland cement). The solution in HC1 is clear, with no suspensions and precipitates, and only slightly yellow.

TABLE 2.

*Some Analyses of Mortars from Wislica*

| *Object* | *Sample Number* | *Element* | *% Sand* | *% $CaCo_3$* | *% Sol.* |
|---|---|---|---|---|---|
| *Early Church* | 221 | wall, middle | 63.9 | 30.1 | 6.0 |
| | | | 65.4 | 30.2 | 4.4 |
| | 22 | wall, middle | 62.1 | 33.7 | 4.2 |
| | | | 68.4 | 27.8 | 3.8 |
| | 18 | wall, top | 68.4 | 26.7 | 4.9 |
| | | | 71.0 | 25.5 | 3.5 |
| | 28 | wall, bottom | 74.8 | 11.6 | 13.7* |
| | | | 79.1 | 7.7 | 13.2 |
| *Romanesque Church* | 9 | northwest wall | 44.9 | 46.0 | 9.1 |
| | | | 43.4 | 45.6 | 11.0 |
| | 124 | northwest wall | 44.9 | 42.2 | 12.9 |
| | | | 44.2 | 40.4 | 15.4 |
| | 123 | southwest wall | 48.4 | 42.5 | 9.1 |
| | | | 44.8 | 46.2 | 9.0 |
| *Gothic Walls* | 122 | outside, reinforcement | 80.8 | 13.7 | 5.5 |
| | | | 82.1 | 13.6 | 4.5 |
| *Collegiate Church* | 119 | inside, south | 71.1 | 17.3 | 11.6 |
| | | | 71.7 | 18.4 | 10.5 |
| | 120 | inside, southwest | 71.6 | 21.2 | 7.2 |
| | | | 70.7 | 20.4 | 8.9 |
| | 108 | inside, east | 74.0 | 16.0 | 10.0 |
| | | | 71.3 | 18.7 | 10.0 |

* The sample from the bottom was strongly corroded, from the top only slightly.

# Metallurgy of Some Ancient Egyptian Medical Instruments

By James Mellichamp U. S. Army Signal Research and Development Laboratory and Martin Levey *
Yale University

## INTRODUCTION

Copper has been known in Egypt for over 6000 years. Many copper objects have been found there belonging to the Badarian period.[1] Copper was not used extensively, however, until the end of the predynastic era (c. 3200 B.C.). The ores for this copper were probably obtained from Sinai or from the eastern desert. By 500 B.C. and later, some of the ancient workings and smelting operations were fairly large, judging by the remaining slag heaps. No doubt,

* The authors would like to thank the University Museum, University of Pennsylvania, for its cooperation in the loan of the Egyptian artifacts. Some of the laboratory work, X-ray diffraction, X-ray fluorescence, and radiographic study, was carried out by W. T. Battis and his research staff at the Central Research Laboratories, American Smelting and Refining Co., South Plainfield, New Jersey. Contributing of his extensive knowledge on spectrographic methods, Charles E. Harvey, Washington State University, also aided in this study in many ways.

One of the authors, Martin Levey, is indebted to the National Institutes of Health for support of this work as part of an overall research program under R. G. 7391.

[1] G. Brunton and G. Caton Thompson, *The Badarian Civilization* (London, 1928), pp. 27, 33.

much of the copper used then was imported from Europe or Asia. Lucas has listed the many ore sites in the neighborhood of Egypt.[2]

Some experts claim that immediately prior to the First Dynasty, the Egyptians had learned all about metallurgy from some invading power. This is highly doubtful. However, prior to 1000 B.C., all the chief useful metals were being worked by Egyptians. This is not too difficult to understand since, with the exception of the tin ores, all of the other necessary minerals were to be found in Egypt. The cupriferous ores were of a readily reducible nature. From the evidence, it is known that they were primarily blue and green carbonates and silicate. Ferruginous and siliceous sands for use as fluxes during ore smelting were abundant. Spoils of war, tribute, and trade accounted for the rest of the copper in the form of imports.

Not much is known of tin in Egypt. It was probably imported in the form of ore or metal. Tin was not in common use in Egypt until the Eighteenth Dynasty (c. 1300 B.C.).[3]

Later on, in the latter half of the first millennium B.C., no doubt, the conquering Persians introduced some of their metallurgical processes since they were skilled in this area. The Egyptians in the Twenty-seventh Dynasty were under the Persians but, in the next three dynasties, they governed themselves with the aid of Greek mercenaries. The Persians then reconquered the country for a short time until the Macedonian invaders, c. 332 B.C., with the aid of the Egyptians, overthrew them. There were

[2] A. Lucas, *Ancient Egyptian Materials and Industries* (London, 1948), pp. 231 ff. *Cf.* J. R. F. Sebelien, *De forhistoriske bronsers.* (Kristiana, 1923).

[3] H. Garland and C. O. Bannister, *Ancient Egyptian Metallurgy* (London, 1927), pp. 28–29; W. M. F. Petrie, *Arts and Crafts of Ancient Egypt* (Edinburgh, 1909), p. 100.

Fig. 1. Egyptian medical instruments.

**Fig. 2. Statuette of the god Osiris.**

then many different forces operating on the material culture of the Egyptians.

The major concern of this investigation is the metallurgy of six Egyptian artifacts. These copper or copper alloy castings, belonging to the latter part of the first millennium B.C., were studied chemically and physically.[4] Four are medical, or embalming, instruments, and for comparison, two Osiris statuettes, one a little smaller than the other, were selected. Dating is uncertain but they are probably from the period 500–300 B.C. The medical instruments are shown in Fig. 1, and the larger Osiris statuette is shown in Fig. 2. The medical instruments consist of two retractors, a tweezers or forceps, and a long-handled, fairly flat spoon or spatula. The latter must have been for solids only. The retractor, which looks like a hook, is 27½ cm. long. The tweezers and spoon are in excellent condition; the other objects show deterioration from corrosion in varying degrees.

Religious statuettes such as the Osiris pieces were mass produced in Egypt. These were cast, either hollow or solid. The Osiris figures which were studied were both solid castings. The casting method generally used was the *cire perdue* process. A beeswax model of the object was first made. This was coated with a material such as clay, then embedded in sand. When heated, the wax ran off, and the metal was poured in. After cooling, the mold was removed. The object was then hammered, filed, or chiseled.[5]

[4] *Cf.* G. B. Phillips, "The Composition of Some Ancient Egyptian Bronzes," *Ancient Egypt,* 1924, 89: Lucas, *op. cit.,* pp. 228ff.

[5] Lucas, *op. cit.,* p. 254; G. Roeder, *Ägyptische bronzewerke* (New York, 1937), pp. 187 ff.; G. A. Wainwright, "Egyptian Bronzemaking," *Antiquity,* 1943, *17,* 96–98, 1944, *18,* 100–102; for Babylonian work, *cf.* M. Levey, *Chemistry and Chemical Technology in Ancient Mesopotamia* (Amsterdam, 1959).

The method of casting the Osiris statuettes is mentioned since this would necessitate a certain composition of copper alloy for a satisfactory result. Pure copper is difficult to use in the manufacture of a solid casting.

## X-RAY DIFFRACTION

One interest in examining these objects was a hope that some feature in the phase relations could be found for use as a time-reckoning index. One possibility is the delta phase of the copper-tin alloy system which is temperature-time dependent. The object with the highest tin content, the smaller Osiris statuette with 4 percent tin, was selected for examination by X-ray diffraction to determine its alloying phases.

The copper-tin alloy system forms a series of electron compounds. The face-centered cubic alpha phase is the most common. $Cu_5Sn$, of the 3/2 type (3 electrons per 2 atoms), is the beta phase having a body-centered cubic structure. The gamma phase is often found in brass and is a complex cubic, body-centered structure. The beta and gamma phases do not exist at lower temperatures. They cannot be preserved by quenching. In fact, quenched alloys make up a group of complex, metastable, transition structures.

The delta phase, formed from gamma and zeta peritectoidally at 592° C, is the 21/13 electron compound. It has the formula $Cu_{31}Sn_8$, containing also the regular gamma structure. The zeta phase is a close-packed hexagonal structure. The epsilon phase is the normal 7/4 electron compound, $Cu_3Sn$, with an orthorhombic structure. The delta phase changes with time, but the transition is apparently not slow enough to be of value in the study of ancient artifacts. After 63 days at 300° C, about 5 percent of the delta in a 2/3 Cu-1/3 Sn alloy is transformed. The process of the eutectoid reaction is that of a pearlitic aggregate.[6] Delta changes to alpha plus epsilon.

In the study of the small Osiris, the copper-base matrix was found to consist of only the alpha phase. In a normal casting with some tin content, some non-equilibrium delta phase is usually present. This may be removed by heat treatment, which is a function of time and temperature, in order to attain equilibrium conditions. Had delta been detected, it would have meant that 2500 years at ambient temperature is not sufficient to restore the equilibrium state in this alloy system. The absence of delta, therefore, permits us to draw no positive conclusions. In other words, although the original casting probably contained delta, either the long period of time or intentional or accidental heat treatment, such as a fire, may have restored equilibrium conditions. Since this object with the highest tin content had no delta phase, it was concluded that there was no possibility of any other of the objects with this phase.

Radiographs revealed some structural weaknesses in several of the objects because of corrosion. However, nothing of value with respect to gross internal structure was found.

## SPECTROCHEMICAL ANALYSIS

The artifacts were analyzed by emission spectroscopy to determine just what elements were alloyed with the copper and what elements appeared in trace quantities.

The artifacts were sampled by taking drillings made in areas least likely to deface or to do damage. Not more

[6] R. W. James, *The Crystalline State* (London, 1939–1953), Vol. II, R. M. Brick, "Phase Diagrams of Copper Alloy Systems," pp. 457, 582.

than a 50-mg. sample was taken from any one piece. The surface portion of the drillings was either discarded or analyzed separately from the interior sampling. A 10-mg. portion of each drilling was weighed into specially designed, high-purity graphite electrodes. The electrodes were placed in the excitation stand of a two-meter grating spectrograph. The sample materials were "burned-to-completion" [7] in a direct current arc.

Fig. 3 shows a small portion, 2800–2900 A, of the analytical spectral region photographed, 2200–4500 A. The spectrograms of the artifacts are located between those of pure copper and standards of the National Bureau of Standards. Several of the analytical spectral lines are indicated at the top of the figure. Most of the results were obtained by visual estimates based on the differences in densities of spectral lines of known values and the corresponding lines of the analyzed material. When chemical values were not available, estimates were based on published sensitivity values for a given spectral line [8] of the elements being determined.

For some of the elements which were present in a sufficient amount and for which several standard values were given, analytical working curves were established for greater accuracy. Working curves were made for Pb, Sn, As, Sb, and Fe by measuring on a comparator-densitometer the densities of a copper internal standard line in each spectrogram and the densities of lines of chemically analyzed elements. The ratios of the copper line to the analytical line were plotted against the known chemical values. The same ratios were made for the unknown material; the chemical values were determined from the curve.

[7] C. E. Harvey, *A Method of Semi-Quantitative Spectrographic Analysis* (Glendale, Calif., 1947).
[8] Ibid.

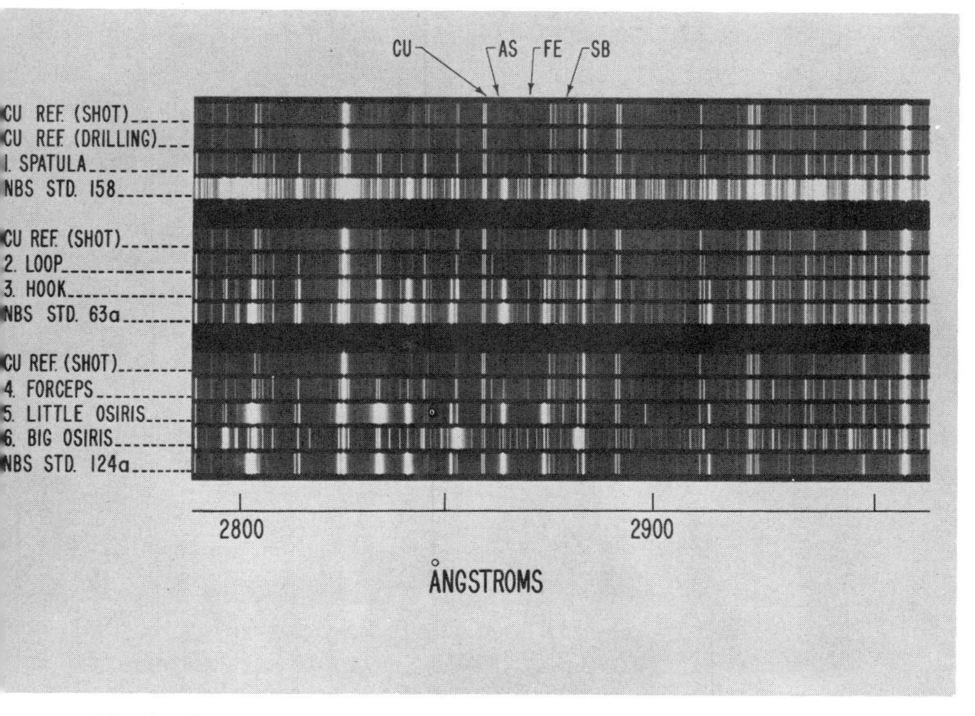


**Fig. 3. Spectrograms used in the spectrochemical analysis of Egyptian artifacts. Some analytical spectral lines are indicated at top.**

An example of a working curve is shown in Fig. 4; this is for tin determinations.

Elements which were present in trace amounts or which had poor spectral sensitivities were determined by a second series of spectrograms. For this analysis, the spectrographic camera was racked after fixed time-periods while the arc was in operation. By this "time study" method, the emitted spectrum of a given element may be caught as the element is preferentially volatilized into the arc column. Thus, the detection limits are increased by as much as a factor of ten over the "burned-to-completion' method employed in the first analysis.

Table I shows the spectrochemical results. In addition to the 22 elements detected and analyzed in most of the artifacts, several additional elements not present in the interior of the materials were detected in the surface samples.

From spectrochemical analysis alone a number of deductions can be made. For example, in the large Osiris and the looped instrument, analyses of the surfaces indicate obvious differences from that of the interior for some elements. This may be due to ambient contaminations from dust, from adhering particles from the original mold, from being hammered, etc. High values for Si, Al, Mg, and Ca indicate contamination from a rock source. The additional presence of K and P in the loop surface suggest granite. The Osiris surface is very high in silicon with also a high content extending into the interior. Possibilities are that the porous surface, which was strongly oxidized, permitted the entrance of sand. When a comparison is made of compositional changes between interior-exterior values of the two artifacts, implications are that the two were made from copper originating from different ore veins.

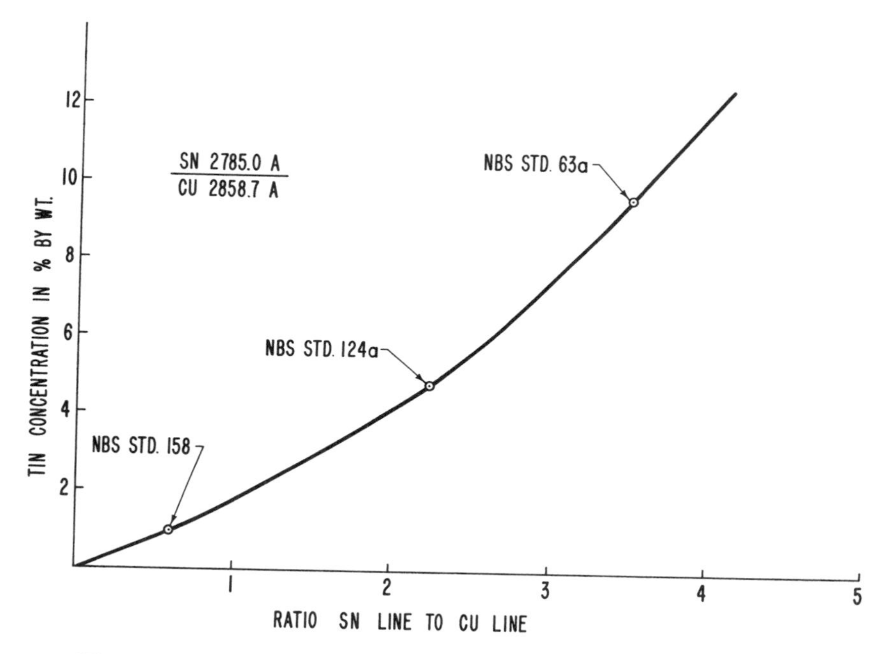


**Fig. 4. Spectrochemical analytical curve used for the determination of tin in Egyptian copper artifacts.**

Spectrochemical analyses show that Sn, Pb, and possibly Zn were intentionally added to copper to form alloys suitable for casting or to be wrought. The presence of arsenic in more than a trace amount indicates that the Egyptians experienced some difficulty in working copper into sheets without first heating to a red heat. With pure copper, in common with other metals of face-centered cubic crystal structure, slip occurs in the close-packed planes [111] in the close-packed [100] directions. With four such planes and three directions in each, there are then twelve possible systems by which slip may occur. This multiplicity of slip systems, here disturbed by arsenic, gives a partial explanation of the capacity for cold work of copper. This may be compared with the structure of zinc and cadmium and their cold-working properties.[9]

The ratio of As/Sb in the artifacts is relatively high, probably indicating that smelting was inefficient to the point where As, among other impurities, could not be properly removed. Further, this ratio is similar for the medical instruments but different for the small Osiris, perhaps implying that the latter was made from different ore material. The high cerium content of the large Osiris also points in this direction. The Co/Ni ratio seems to be fairly constant for the medical instruments but is much lower in the large Osiris. The meaning of this in terms of the ore sources is uncertain.

## CONCLUSIONS

The application of modern technology to ancient bronzes revealed information about the state of metallurgy at the

[9] D. W. Wakeman, "The Physical Chemistry of Copper," in A. Butts, *Copper, the Science and Technology of the Metal, its Alloys and Compounds* (New York, 1954), p. 418.

TABLE I.

Spectrochemical Analysis of Ancient Egyptian Bronzes *

| | 1 | 2a | 2b | 3 | 4 | 5 | 6a | 6b |
|---|---|---|---|---|---|---|---|---|
| *Element* | *Spatula* | *Loop* | *Loop Surface* | *Hook* | *Forceps* | *Little Osiris* | *Big Osiris* | *Big Osiris Surface* |
| *Cu* | MAJOR CONSTITUENT | | | | | | | |
| *Pb* | 0.02 | 0.05 | 0.05 | 1. | 0.02 | 10. | 2. | 1. |
| *Sn* | 0.004 | 0.005 | 0.005 | 1. | 0.01 | 4. | 0.3 | 0.02 |
| *Zn* | 0.005 | 0.05 | 0.004 | 0.05 | 0.002 | 0.2 | 1. | 0.5 |
| *As* | 0.1 | 0.3 | 0.3 | 0.1 | 0.1 | 0.1 | — | — |
| *Sb* | 0.03 | 0.04 | 0.04 | 0.08 | 0.04 | 0.1 | — | — |
| *Fe* | 0.4 | 0.3 | 0.8 | 0.6 | 0.3 | 0.1 | 1. | 2. |
| *Si* | 0.003 | 0.004 | 0.2 | 0.08 | 0.01 | 0.1 | 1. | 10. |
| *Al* | 0.003 | 0.004 | 0.2 | 0.01 | 0.003 | 0.02 | 1. | 1. |
| *Mg* | 0.002 | 0.002 | 0.07 | 0.01 | 0.003 | 0.01 | 0.2 | 1. |
| *Ca* | 0.005 | 0.005 | 0.05 | 0.02 | 0.002 | 0.02 | 1. | 2. |
| *B* | 0.0008 | 0.0006 | 0.006 | 0.002 | 0.001 | 0.0006 | 0.0004 | 0.2 |
| *Na* | — | — | 0.3 | — | — | — | — | — |
| *P* | — | — | 0.8 | — | — | — | — | — |
| *K* | — | — | 0.3 | — | — | — | — | — |
| *Ti* | 0.0008 | 0.002 | 0.005 | — | 0.002 | 0.01 | 0.05 | 0.02 |
| *V* | — | — | 0.002 | 0.001 | — | — | — | — |
| *Cr* | 0.0002 | 0.0009 | 0.001 | 0.0001 | 0.0008 | 0.0007 | 0.003 | 0.004 |
| *Mn* | 0.002 | 0.003 | 0.01 | 0.005 | 0.002 | 0.004 | 0.02 | 0.02 |
| *Co* | 0.02 | 0.02 | 0.02 | 0.03 | 0.07 | 0.02 | 0.003 | 0.004 |
| *Ni* | 0.08 | 0.06 | 0.05 | 0.06 | 0.08 | 0.06 | 0.02 | 0.02 |
| *Zr* | — | — | 0.002 | — | — | 0.003 | 0.03 | 0.01 |
| *Mo* | — | — | — | — | — | — | — | 0.01 |
| *Ag* | 0.005 | 0.01 | 0.02 | 0.02 | 0.01 | 0.04 | 0.001 | 0.001 |
| *In* | — | — | 0.003 | 0.003 | — | — | — | — |
| *Ce* | — | — | — | — | — | — | 0.09 | 0.05 |
| *Au* | 0.006 | 0.01 | 0.01 | 0.07 | 0.008 | 0.04 | — | 0.01 |
| *Sr* | — | — | — | — | — | — | 0.01 | 0.01 |

* — = Not Detected. Results are in % by total weight.

time the objects were made. While only a few artifacts were examined at this time by several techniques, some information was obtained, most of it for future use. No doubt, the ancient Egyptians had master coppersmiths who could make the most out of the materials and techniques they understood. However, they were somewhat limited as compared with other ancient cultures. Spectrochemical analysis showed by the presence of a large number of trace impurities that they were able to refine copper only to some extent and were somewhat handicapped in the cold working of copper. By studying differences in ratios of impurity elements, some differentiations of ore sources could be made. Surface analysis as compared with internal analysis gave something of the history of the objects. The hope that some feature in the phase relationship of the copper-tin alloy, such as the presence of the delta phase, could be used as a time index was not realized.

# Die Untersuchung alter eiserner Fundstücke und die dazu verwendeten Verfahren

By F. K. Naumann Max-Planck-Institut
für Eisenforschung, Düsseldorf

(Inhalt: Die Verfahren und der Arbeitsgang, die bei der Untersuchung alter Fundstücke zweckmäßig anzuwenden sind, werden beschrieben und an einigen Beispielen erläutert.)

Wenn wir heute metallische Fundstücke aus alter Zeit nach metallkundlichen Verfahren untersuchen, so verfolgen wir damit den Zweck, Einblick in die Verfahren ihrer metallurgischen und technologischen Herstellung und damit in den Stand der Technik jener Zeit zu gewinnen, in der sie hergestellt worden sind. Solche Untersuchungen können aber auch dem Archäologen bei der Bestimmung von Zeit und Ort der Herstellung eines Fundstückes behilflich sein. In der Erkenntnis, daß eine Zusammenarbeit allen Teilen dienlich ist, haben sich Metallurgen, Archäologen, Geschichts- und Sprachforscher in dem Geschichts- ausschuß des Vereins Deutscher Eisenhüttenleute unter Leitung von Prof. E. H. Schulz zusammengefunden. In gemeinsamer

Arbeit wurden Erfahrungen gesammelt, die zur Aufstellung allgemeiner Richtlinien für das zweckmäßige Vorgehen bei solchen Untersuchungen geführt haben.[1]

Für die Untersuchung alter Fundstücke können physikalische, chemische, metallographische und mechanische Prüfverfahren eingesetzt werden. Die besonderen Bedingungen, die bei solchen Untersuchungen vorgegeben sind, machen aber oft die Anwendung von Sonderverfahren und ein Vorgehen in anderer Reihenfolge zweckmäßig, als sie bei gewöhnlichen Werkstoffuntersuchungen üblich sind. Diese Bedingungen sind gekennzeichnet durch den meist heterogenen Aufbau des alten Schweißeisens aus Luppen verschiedener Zusammensetzung und durch die vielfach erhobene Forderung, die Fundgegenstände möglichst wenig oder gar nicht zu beschädigen. Wie unter solchen Bedingungen vorzugehen ist, soll an einigen Beispielen erläutert werden.

## 1. Zerstörungsfreie Prüfung

Für den Fall, daß wegen des hohen Wertes oder einmaligen Vorkommens eines Fundstückes eine Zerstörung völlig ausgeschlossen ist, lassen sich manchmal schon durch eine zerstörungsfreie Prüfung nach einem der bekannten Verfahren Erkenntnisse über den Aufbau oder die Herstellung des Gegenstandes gewinnen. Aus der Form eines in Luristan (Nordsyrien) gefundenen Kurzschwertes aus dem 7. oder 8. Jahrhundert v. Chr. (Bild 1)[2] hatte ein Archäologe geschlossen, daß es durch Gießen hergestellt sein müßte. Da es bisher als sicher galt, daß zwar die

[1] *Arch. Eisenhüttenwes.* 34 (1963) . 961/63.

[2] F. K. Naumann: Bericht Nr. 19 des Geschichtsausschusses des Vereins Deutscher Eisenhüttenleute, *Arch. Eisenhüttenwes. 28* (1957) S. 575/81

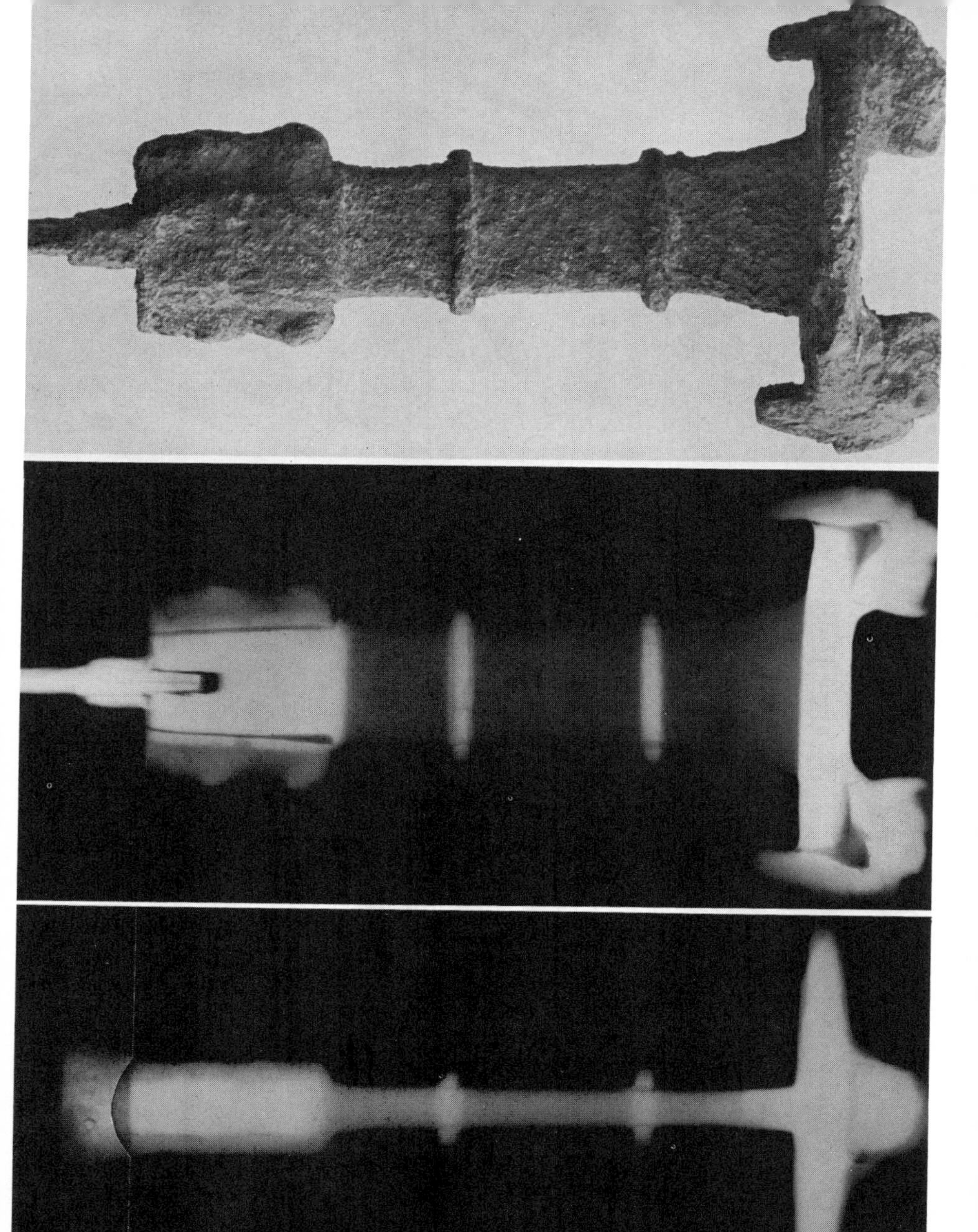

**Bilder 1-3. Griffteil eines luristanischen Kurzschwertes aus dem 7. oder 8. Jahrhundert v. Chr.**
**Bild 1. Seitenansicht.**
**Bild 2. Durchstrahlung von der Seite.**
**Bild 3. Durchstrahlung von der hohen Kante.**

Reduktion des Eisens im festen Zustand und die Verarbeitung des Schweißeisens ihren Ursprung in Vorderasien hat, daß aber die Erzeugung des Gußeisens und die Technik des Eisengießens zuerst in China entwickelt worden sind, hätte diese Auffassung, wenn sie berechtigt gewesen wäre, unsere bisherige Anschauung über die Herkunft des Gußeisens in Frage gestellt. Mit Hilfe einer Durchstrahlung konnte ein Einblick in den Aufbau des Schwertes gewonnen werden. Als Strahlungsquellen wurden für die dünneren Querschnitte des Griffes das Iridium-Isotop Ir 192 und für den dickeren Knauf ein Betatron mit 15 Me V Strahlungsenergie verwendet. In den Bildern 2 und 3 sind Durchstrahlungsaufnahmen des Griffes von zwei Seiten aus wiedergegeben. Darin wird sichtbar, daß die Klinge, die am oberen Ende durch zwei aufgenietete Laschen verstärkt ist, in einen Schlitz des Heftes eingelassen und mit diesem vernietet ist, daß die Figuren des Heftes nicht Bestandteile des Griffes, sondern mit diesem in einer nicht genau erkennbaren Weise verbunden sind, und daß die beiden Wulste, die den Griff unterteilen, aus Drähten bestehen, die um den Griff herumgewunden und mit den Enden stumpf zusammengestoßen sind. Wie die Durchstrahlung des Knaufs, Bild 4, zeigte, ist der Griff in einen Schlitz der Knaufplatte eingelassen, und die bärtigen Männerköpfe sind in eine Nut der Platte eingeschoben. Das Schwert besteht also außer den drei Nieten aus zehn Einzelteilen. Dadurch wird die Annahme, daß der ganze Griffteil des Schwertes aus einem einzigen Gußstück bestünde, bereits widerlegt. Durch eine metallographische Untersuchung der Figuren an Heft und Knauf wurde festgestellt, daß auch diese nicht aus Gußeisen, sondern aus Schweißstahl hergestellt sind. Dazu wird später noch etwas zu sagen sein.

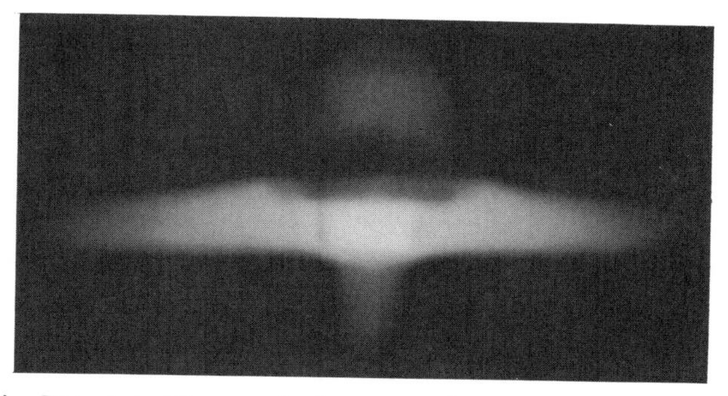

**Bild 4. Durchstrahlungsaufnahme vom knauf eines luristanischen Kurzschwertes, 1:1.**

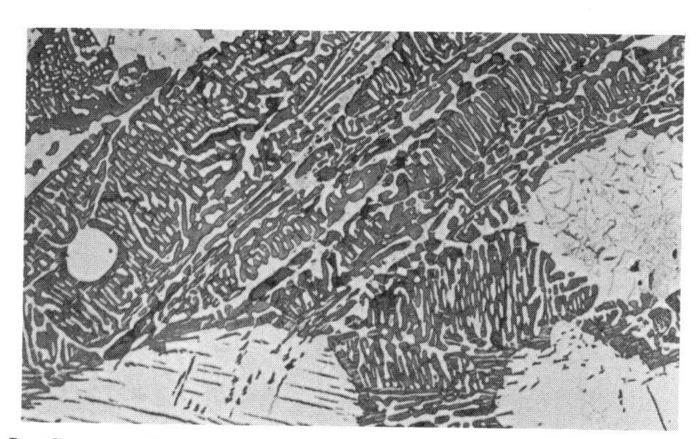

**Bild 5. Spuren des ursprünglichen Roheisengefüges im Rost einer Ofensau aus dem 1. Jahrhundert v. Chr., 200:1.**

Die Durchstrahlung kann auch über den inneren Aufbau des Metalles Aufschluß geben. So wurde z.B. die Technik des Damaszierens von Klingen mit ihrer Hilfe untersucht.[3]

[3] J. Driehaus: Unveröffentlichter Vortrag auf der 19. Sitzung des Geschichtsausschusses des Vereins Deutscher Eisenhüttenleute am 20.6.1961.

## 2. Entrosten

Alte Fundstücke sind meistens mit einer dicken Rostschicht überzogen. Sie sollte nicht ohne Bedenken abgelöst werden; denn manchmal kann auch die metallographische Untersuchung des Rostes noch Aufschlüsse über das ursprüngliche Gefüge geben. Bild 5 gibt z.B. das Gefüge eines ursprünglich metallischen Stückes wieder, das in einer Ofensau aus dem 1. Jahrhundert v.Chr. eingeschlossen war und im Laufe der Zeit völlig verrostet ist. An dem stehengebliebenen Graphit und den Spuren des Ledeburit-Eutektikums wird kenntlich, daß es Roheisen war, welches wahrscheinlich unbeabsichtigt entstanden und mit der Schlacke in den Herd getropft ist. Ein unüberlegtes Entrosten hätte hier zur völligen Zerstörung des Stückes geführt. Aber auch durch oberflächliches Entrosten könnten Schichten abgelöst werden, die für den Aufbau des Stückes kennzeichnend sind. Wenn es nur darauf ankommt, die äußere Form zu erhalten, kann der Rost mit Hilfe eines wasserstoff- oder kohlenoxydhaltigen Gases reduziert werden.

## 3. Metallographische Untersuchung

Weil alte Fundstücke in der Regel aus Schweißeisen bestehen, das im Rennfeuer reduziert und aus mehreren oder vielen Luppen verschiedener Zusammensetzung zusammengeschmiedet ist, sollte man sich vor der Entnahme von Analysenspänen erst durch eine zunächst makroskopische metallographische Untersuchung von dem Gefügeaufbau des Stückes überzeugen. Wenn es erlaubt ist, das Stück ganz oder wenigstens zur Hälfte zu zerstören, wird zweckmäßig ein Schnitt durch das ganze Stück gemacht.

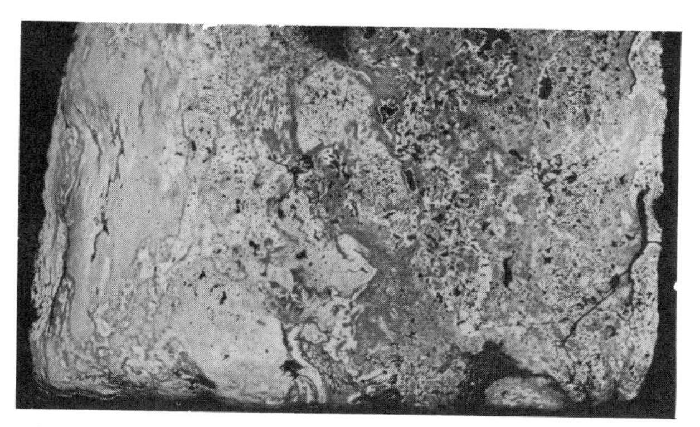

**Bilder 6 und 7. Querschliff durch die Längshälfte eines Spitzbarrens aus Gebiet der oberen Donau.**

**Bild 6. Mit alkohol. Salpetersäure geätzt, 1:1. Die kohlenstoffarmen erscheinen hell, die kohlenstoffreichen dunkel.**

**Bild 7. Nach Oberhoffer geätzt, 1:1. Die phosphorreichen Stellen erscheinen hell, die phosphorärmeren dunkler.**

Der Schnitt kann parallel oder quer zur Schmiederichtung geführt werden. Ein Längsschliff bietet den Vorteil, daß daran auch der Verschmiedungsgrad und der Faserverlauf erkannt werden können. Aus einer der beiden Längshälften sollten zusätzlich Querschliffe entnommen werden. Der Aufbau des Stückes wird häufig schon am ungeätzten Schliff erkennbar, weil die einzelnen Schichten durch unverschweißte Stellen und durch Zeilen von Schlackeneinschlüssen voneinander getrennt sind. Durch makroskopische Ätzung wird man meistens feststellen, daß sich diese Lagen stark im Kohlenstoff—und Phosphorgehalt unterscheiden. Dies soll an der Untersuchung eines im Gebiet der oberen Donau gefundenen Spitzbarrens erläutert werden.[4] Spitzbarren sind vierkantige Eisenknüppel von 50 bis 100 mm Dicke, die nach beiden Enden zu spitz ausgeschmiedet sind. Sie wurden wahrscheinlich von den Schmieden der Latène-Zeit als Vormaterial benutzt. Die Bilder 6 und 7 zeigen einen in verschiedener Weise geätzten Querschliff durch eine Längshälfte des Barrens. Die kohlenstoffarmen Zonen, die im wesentlichen auf der linken Seite des Querschnitts liegen, sind zugleich phosphorreich. Das hängt mit der Abschnürung des $\gamma$-Gebietes durch Phosphor und der geringeren Löslichkeit des Kohlenstoffs in dem durch Phosphor stabilisierten Ferrit zusammen. Der Kohlenstoffgehalt schwankt in demselben Stück von etwa 0,02 bis 1,5 % (Bilder 8 und 9). Der hohe Phosphorgehalt ist im Gefüge auch an den breiten Ferrithöfen am Rande der ehemaligen Austenitkristallite und an einer Gelbfärbung des nicht umgewandelten Ferrits beim Ätzen in alkoholischer Pikrinsäure zu erkennen, (Bild 10). Das Gefüge

[4] W. Rädeker und F. K. Naumann: Bericht Nr. 43 des Geschichtsausschusses des Vereins Deutscher Eisenhüttenleute, Arch. Eisenhüttenwes. *32* (1961) S. 587/95.

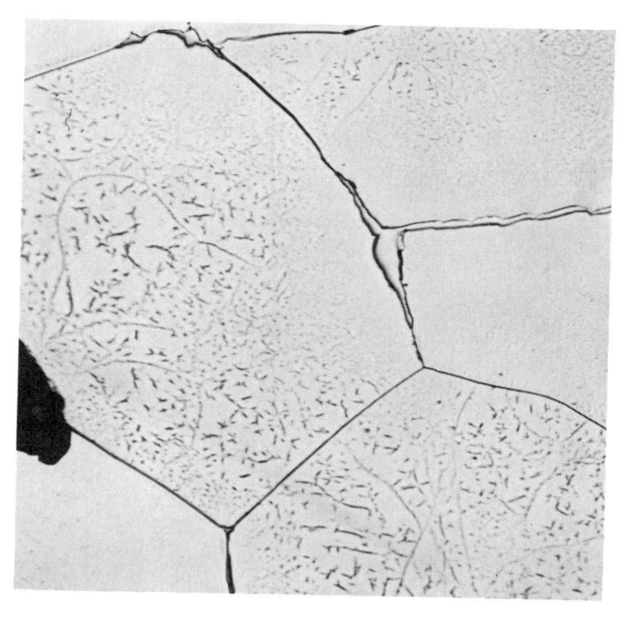

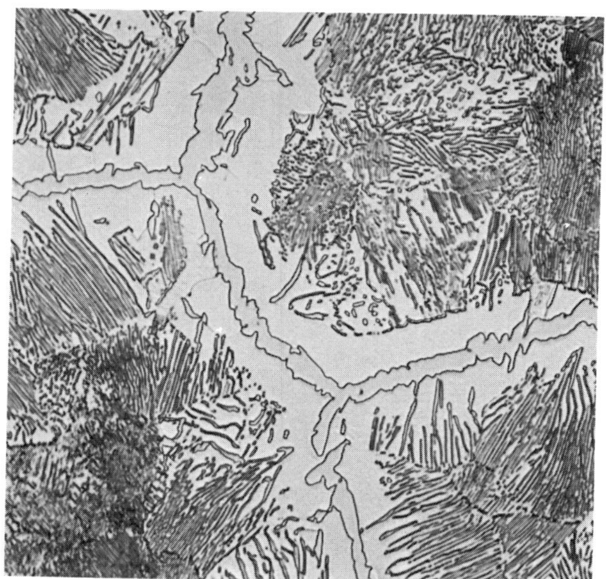

**Bilder 8 und 9. Gefüge in einem Spitzbarren. Querschliff, Atzung: alkohol. Salpetersäure, 500:1.**

**Bild 8. Kohlenstoffarme Schicht.**

**Bild 9. Kohlenstoffreiche Schicht.**

ist anomal ausgebildet (Bild 9), weil im Rennfeuer nur wenig Silizium und Mangan reduziert werden kann.

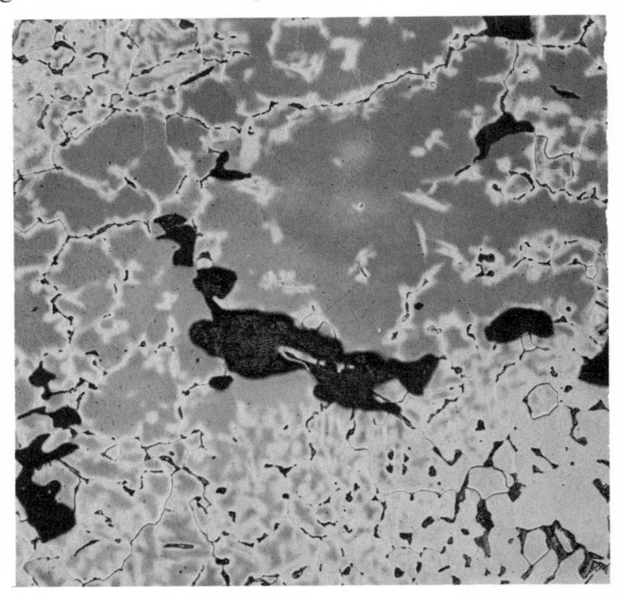

**Bild 10. Gefüge in der phosphorreichen Schicht eines Spitzbarrens. Querschliff, Ätzung: alkohol. Pikrinsäure, 200:1.**

Auch den eingeschlossenen Schlacken sollte bei der metallographischen Untersuchung Aufmerksamkeit gewidmet werden. Sie bestehen in der Regel aus dendritisch ausgeschiedenem Wüstit mit plattenförmigem Fayalith (2 FeO . $SiO_2$) in einer kieselsäurereichen, eutektischen oder glasigen Grundmasse (Bild 11): Für die Identifizierung dieser Phasen hat sich die Untersuchung im polarisierten Licht bewährt. Beim Drehen zwischen gekreuzten Nicols bleibt der undurchsichtige kubische Wüstit immer dunkel, während der durchsichtige rhombische Fayalith je Umdre-

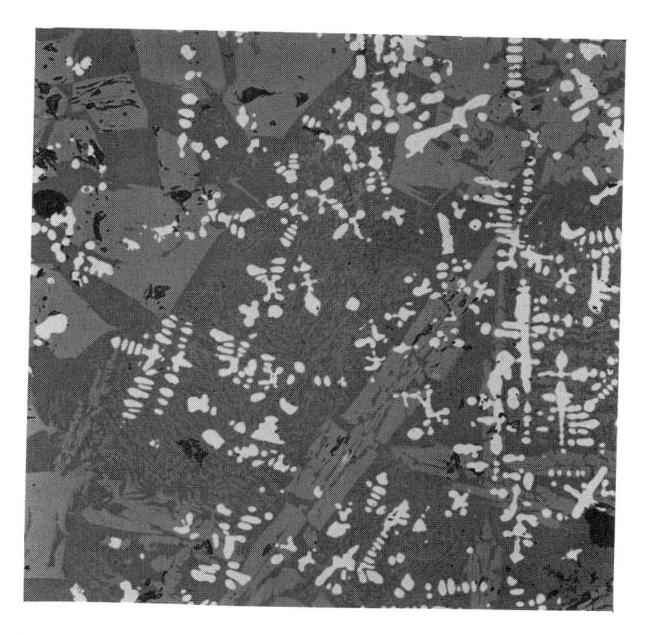

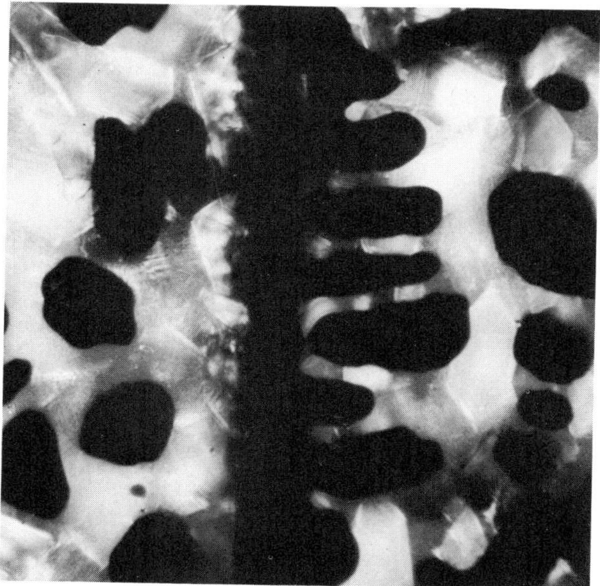

**Bilder 11 und 12. Gefüge eines Schlackeneischlusses aus der kohlenstoffarmen Zone eines Spitzbarrens.**

**Bild 11. Normal beleuchtet, 100:1.**

**Bild 12. Im polarisierten Licht zwischen gekreuzten Nicols, 1000:1.**

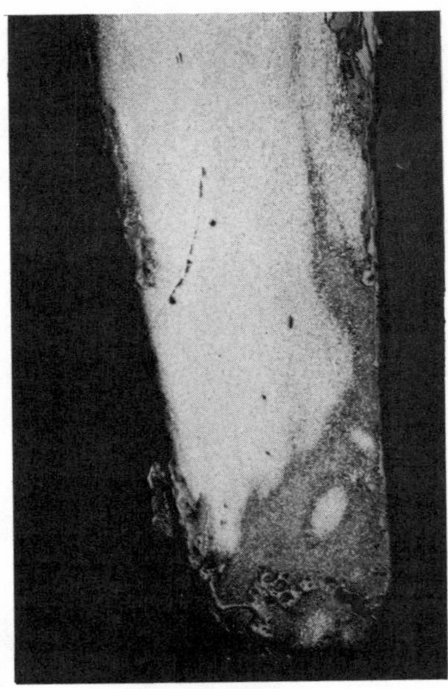

**Bild 13. Längschliff durch die Spitze eines Nagels, der auf dem Magdalensberg in Kärnten gefunden wurde. Atzung: alkohol. Salpetersäure, 10:1.**

hung viermal aufleuchtet und wieder dunkler wird (Bild 12). In den Einschlüssen der kohlenstoffreichen Schichten fehlt meistens der Wüstit und manchmal auch der Fayalith.

Kohlenstoffreiche und kohlenstoffarme Zonen sind in den Stücken meistens regellos verteilt, ein Zeichen dafür, daß die alten Schmiede die Erzeugung eines einheitlichen Werkstoffes im allgemeinen nicht in der Hand gehabt haben. Nur wenige Ausnahmen sind dem Verfasser bekanntgeworden: im Raume des alten Noricum, dem heutigen Gebiet der österreichischen Alpenländer Kärnten und Steiermark, wurden überwiegend Werkzeuge aus einem kohlenstoffreichen Stahl gefunden. Der norische Stahl war schon im Altertum wegen seiner hohen Härte geschätzt. Dort scheint man demnach über Verfahren verfügt zu haben, das aus den phosphorarmen Erzen der einheimischen Lagerstätten reduzierte Eisen im Zusammenhang mit der Reduktion oder nachträglich so weit aufzukohlen, daß ein einheitlicher Werkzeugstahl entstand. Auch das Verstählen, d.i. das Aufschweißen kohlenstoffreicher Luppen auf kohlenstoffärmeren Grundwerkstoff an Stellen, die eine hohe Härte oder Verschleißfestigkeit haben sollten, scheint dort bekannt gewesen zu sein. Als Beispiel wird in Bild 13 das Gefüge eines Nagels mit kohlenstoffreicherer Spitze gezeigt, der auf dem Magdalensberg in Kärnten ausgegraben wurde.[5]

Die metallographische Untersuchung gibt ferner Aufschluß über die Wärmebehandlung der Stücke. In den keltischen und römischen Siedlungen auf dem Magdalensberg wurden auch Werkzeuge gefunden, die an den Schneiden oder Spitzen durch Abschrecken gehärtet waren. In den Bildern 14 und 15 sind der Längsschliff durch ein solches Werkzeug und das martensitische Gefüge in seiner gehärteten Spitze wiedergegeben. Auch eine kunstvolle Kombination von gehärtetem mit nicht härtbarem Stahl ist schon im Altertum durchgeführt worden, wie die Untersuchung römischer Schwerter, die bei Nydam am Alsensund gefunden wurden, gezeigt hat.[6] Dagegen ist dem Verfasser kein Fall einer Einsatzhärtung bekanntgeworden.

Wenn eine Zerstörung oder Beschädigung des zu unter-

[5] F. K. Naumann: Vortrag auf der 21. Sitzung des Geschichtsausschusses des Vereins Deutscher Eisenhüttenleute am 5.6.1962, Arch. Eisenhüttenwes 35 (1964). 495/502.

[6] E. Schürmann: Bericht Nr. 30 des Geschichtsausschusses des Vereins Deutscher Eisenhüttenleute, Arch. Eisenhüttenwes. *30* (1959) S. 121/26.

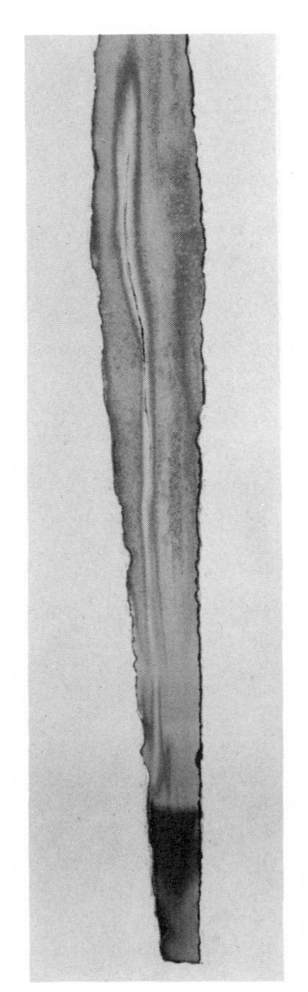

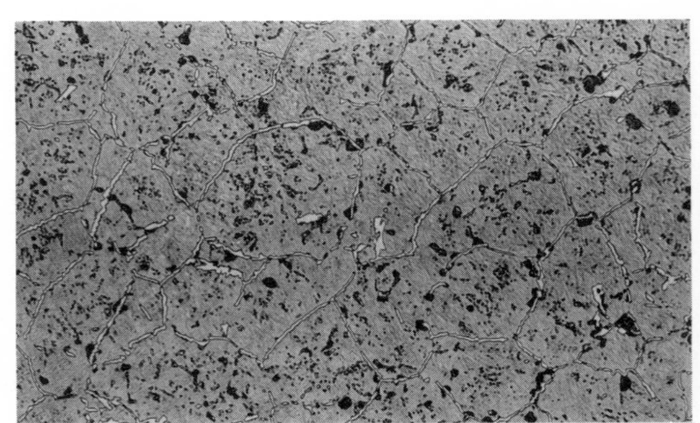

**Bilder 14 und 15. Doppelstachel vom Magdalensberg. Atzung: alkohol. Salpetersäure.**

**Bild 14. Längschliff.**

**Bild 15. Gefüge in der Spitze, 200:1.**

suchenden Stückes vermieden werden soll, muß auch der Metallograph mit Sonderverfahren arbeiten. Oberflächenanschliffe ergeben wegen der Randentkohlung und Korrosion meistens kein zutreffendes oder vollständiges Bild des Gefügeaufbaues. Im Falle des erwähnten luristanischen Kurzschwertes konnte man sich, wie in den Bildern 16 und

**Bild 16. Schliffstelle am Heft eines luristanischen Kurzschwertes, 1:1.**

17 gezeigt wird, damit helfen, daß die dicke Rostschicht mit einem schnellaufenden kegelstumpfförmigen Bohrer von 1 mm ∅ durchbohrt und die Stirnfläche des Bohrlochs mit einem Holzstäbchen unter Zusatz von aufgeschlämmtem Schmirgel und Diamantpulver oder Tonerde geschliffen und poliert wurde. Auf diese Weise konnte, ohne daß das Schwert wesentlich beschädigt wurde, festgestellt werden, daß auch das Heft und die Figuren des Knaufs aus kohlenstoffarmem Schweißeisen und nicht aus Gußeisen bestanden.

**Bild 17. Gefüge im Heft eines luristanischen Kurzschwertes, 200:1. Atzung: alkohol. Salpetersäure.**

## 4. Härteprüfung

Als Ergänzung zur metallographischen Untersuchung kann die Härteprüfung herangezogen werden. Sie gibt ein gutes Bild über die Anordnung kohlenstoffreicher und kohlenstoffarmer Schichten. Als Beispiel ist in Bild 18 der geätzte Querschliff durch eine Lanzenspitze aus dem 7. nachchristlichen Jahrhundert wiedergegeben und in Bild 19 das Ergebnis der Härtemessungen über die Dicke des Blattes an den in Bild 18 gekennzeichneten Stellen. Der Härteverlauf folgt genau dem Wechsel der Schichten; die Härte HV 0,1 beträgt in den kohlenstoffreichen Zonen mit 0,6 bis 0,8 % C 200 bis 260 kg/mm² und in den kohlenstoffarmen Zonen mit rund 0,1 % C 110 bis 130 kg/mm². Für die Härtebestimmung wird in solchen Fällen am besten ein Kleinlasthärteprüfer mit 100 bis 1000 g Prüflast verwendet.

## 5. Chemische Untersuchung

Erst wenn der Aufbau des zu untersuchenden Stückes durch eine metallographische Untersuchung bekannt ist, kann eine gezielte chemische Analyse angesetzt werden. Eine Analyse aus Spänen, die ohne die beschriebenen Voruntersuchungen entnommen wären, könnte nur Zufalls- oder Durchschnittswerte ergeben, die einen unvollständigen oder irreführenden Eindruck von der Zusammensetzung des Stückes vermitteln würden.

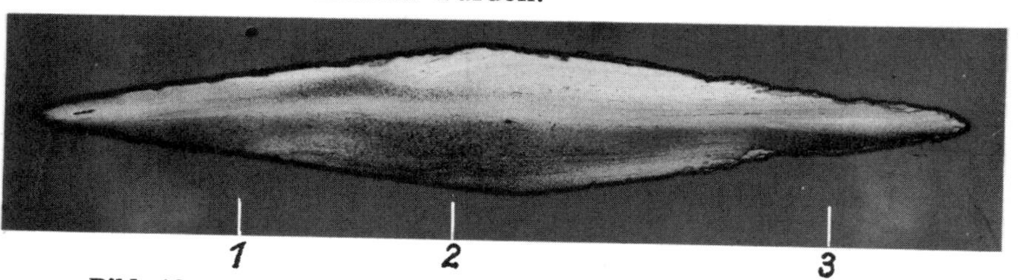


**Bild 18. Querschliff durch eine Langenspitze aus dem 7. Jahrhundert. Atzung: alkohol, Salpetersäure, 2,5:1. 1,2,3 = Lage der Messreihen für die Härteprüfung.**

Es wurde schon erwähnt, daß Kohlenstoff und Phosphor in beträchtlicher Menge in dem Eisen alter Fundstücke vorkommen können. Der Schwefelgehalt ist dagegen im

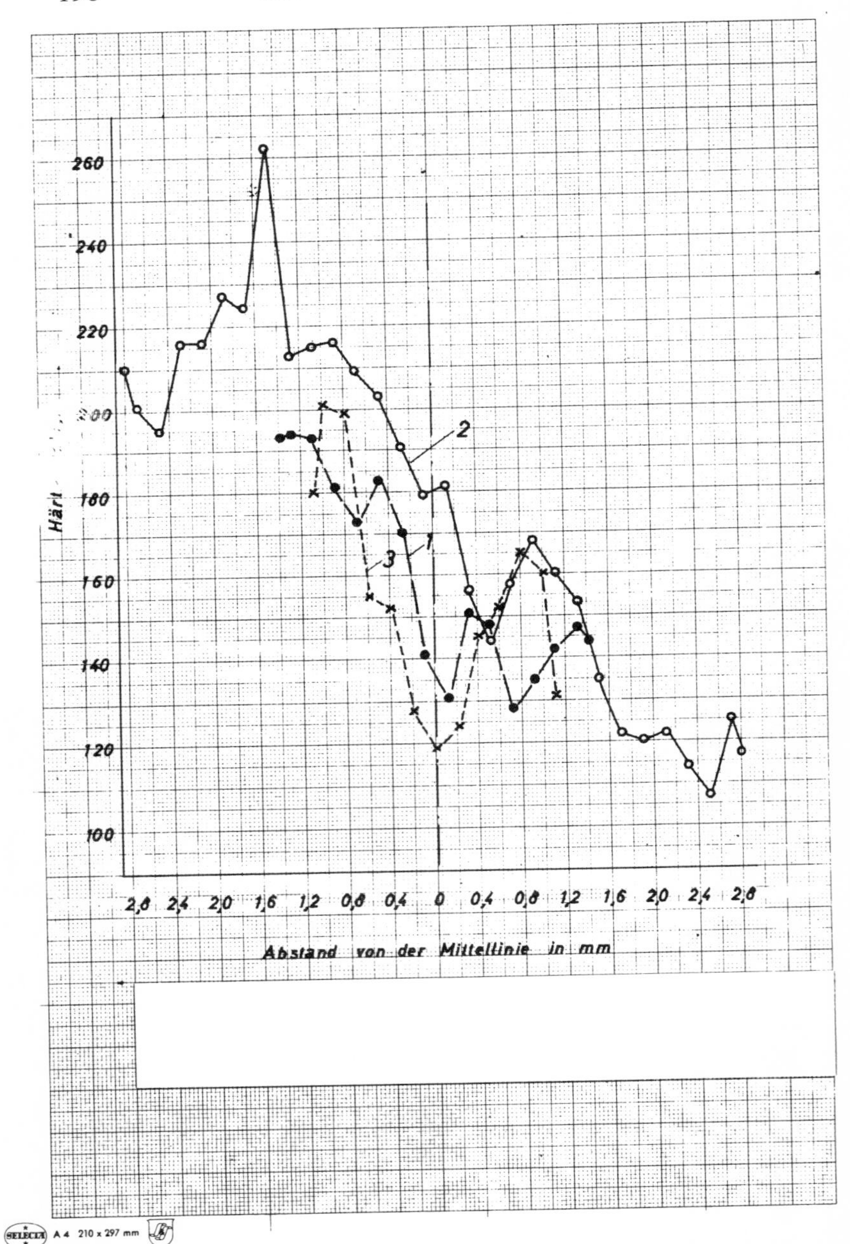


**Bild 19. Härteverlauf über die Blattdicke der Langenspitze.**

Rennfeuereisen wegen der Schwefelarmut der als Reduktionsmittel verwendeten Holzkohle sehr gering. Auch Stickstoff wird bei der Reduktion im festen Zustand nicht in nennenswerter Menge aufgenommen. Trotzdem sollte der Stickstoffgehalt, besonders in kohlenstoffreichen Zonen, mitbestimmt werden, um festzustellen, ob der Kohlenstoff nicht durch nachträgliches Aufkohlen in organischen Stoffen eingebracht worden ist. Nach Möglichkeit sollte auch auf Kupfer, Nickel und Arsen untersucht werden, da ihre Anwesenheit einen Hinweis auf die Herkunft der verwendeten Erze und damit des Fundstückes geben kann. Besondere Vorsicht ist bei der Bestimmung von Silizium und Mangan geboten. Diese schwer reduzierbaren Metalle sind im Rennfeuereisen nur in sehr geringen Mengen vorhanden. Sie finden sich aber in beträchtlicher Menge in den eingeschlossenen Schlacken. Es ist deshalb erforderlich, den metallischen und den oxydischen Teil getrennt zu analysieren. Zur Trennung wird in solchen Mitteln gelöst, die nur das Metall auflösen. Am besten eignet sich die elektrochemische Isolierung nach P. Klinger und W. Koch. Wenn trotz dieser Vorsichts-maßnahmen in dem metallischen Teil Silizium- und Mangangehalte in der Größenordnung von Zehntel Prozenten gefunden werden, ist der Schluß berechtigt, daß das Eisen nicht aus einem normalen Rennfeuerprozeß, sondern aus einem Verfahren mit stärkerer Reduktion hervorgegangen ist.

In den Schlacken sollten Eisen(II)- und Eisen(III)-Oxyd, Mangan(II)-Oxyd, Kieselsäure, Phosphorsäure, Tonerde, Kalk und Magnesia, gegebenenfalls auch Titandioxyd bestimmt werden. Rennfeuerschlacken sind immer sehr eisenreich, weil nur solche Schlacken bei den in diesem Verfahren erreichbaren Temperaturen flüssig sind. Die Rennfeuerschlacken des westdeutschen Siegerlandes und

**Bild 20. Einschluss von Holzkohle in einer Herdschlacke vom Magdalensberg, 100:1.**

der österreichischen Alpen enthalten, soweit sie bei der Verhüttung der dort anstehenden Siderite oder deren Verwitterungsprodukte angefallen sind, oft beträchtliche Mengen Mangan(II)-Oxyd, und in den Schlacken aus Kärnten und der Steiermark wird meistens Titandioxyd aus dem Ilmenit der dortigen Erze gefunden. Solche Untersuchungen können demnach bei der Einordnung von Fundstücken unbekannter Herkunft helfen. Aus der Beziehung des (Fe0 + Mn0)-Gehaltes zum $SiO_2$-Gehalt der Schlacke kann nach E. Schürmann [7] geschlossen werden, ob das Eisen in einem Rennfeuer einfacher Bauart oder in

[7] E. Schürmann: Bericht Nr. 24 des Geschichtsausschusses des Vereins Deutscher Eisenhüttenleute, Stahl und Eisen 78 (1958) S.1297/1308.

einem Ofen mit stärkerer Eisenreduktion hergestellt worden ist.

In Schlacken und Ofensauen findet man manchmal auch noch Reste von Holzkohle (Bild 20). An solchen Stücken, aber auch an kohlenstoffreichem Eisen, müßte auch eine Feststellung ihres Alters durch die Bestimmung des Kohlenstoffisotops C 14 möglich sein.

Da bei solchen gezielten chemischen Untersuchungen meist nur geringe Substanzmengen zur Verfügung stehen, müssen zur Bestimmung der Elemente vielfach mikrochemische Verfahren angewendet werden. An dem erwähnten Spitzbarren aus dem Gebiet der oberen Donau wurden im Metall und in den elektrochemisch isolierten Schlackeneinschlüssen die folgenden Gehalte festgestellt:

| *Zone* | *Metallischer Teil* | | | | |
|---|---|---|---|---|---|
| | *C* % | *Si* % | *Mn* % | *P* % | *S* % |
| *C-arm* | 0,072 | 0,015 | 0,018 | 0,45 | Sp. |
| *C-reich* | 0,652 | 0,014 | 0,030 | 0,18 | Sp. |

| *Zone* | *Schlackeneinschlüsse* | | | | | |
|---|---|---|---|---|---|---|
| | *FeO* % | *$SiO_2$* % | *$Al_2O_3$* % | *CaO* % | *MgO* % | *MnO $P_2O_5$ $TiO_2$* % |
| *C-arm* | 71,5 | 19 | 5 | 3,5 | 1 | 0 |
| *C-reich* | 65 | 25 | 5 | 4 | 1 | 0 |

Wie bereits metallographisch ermittelt wurde, ist im kohlenstoffarmen Teil der Phosphorgehalt des Metalls und der Eisengehalt der Schlackeneinschlüsse höher als im kohlenstoffreichen Teil. Bemerkenswert ist, daß die Unterschiede im Kohlenstoffgehalt längst nicht so groß sind, wie sie bei der Gefügeuntersuchung in Erscheinung traten (vgl. die Bilder 8 und 9). Das beweist, daß die chemische Untersuchung auch dann noch Durchschnittswerte aus einem Bereich mit heterogener Zusammensetzung liefert, wenn man sich bemüht, das Spänevolumen möglichst klein zu halten.

**Bild 21. Gefüge im Blatt einer Langenspitze aus dem 7. Jahrhundert n. Chr. Längschliff, Ätzung: alkohol, Pikrinsäure, 100:1.**

Wenn ein Element auf kleinstem Raum angereichert ist, aber nicht als besondere Phase vorliegt, so daß es chemisch oder mechanisch isoliert werden könnte, müssen andere Verfahren zu seiner Bestimmung angewendet werden, wie die Anregung seiner Eigenstrahlung in der Elektronen-Mikrosonde oder die Röntgenfluoreszenzanalyse. Die erwähnte Lanzenspitze aus dem 7. Jahrhundert (vgl. Bild 18) zeigte in ihrem kohlenstoffreichen Teil ein perlitisches Gefüge von anomaler Ausbildung. Nur in einzelnen Streifen von 0,05 bis 0,2 mm Breite erschien der Perlit feinstreifig und damit normal (Bild 21). Es wurde vermutet, daß in diesen Streifen der Mangangehalt höher als in dem anomalen Grundgefüge ist, weil Mangan die Unterkühlung der $\gamma$ $\alpha$-Umwandlung und damit die Bildung eines feinstreifigen Perlits begünstigt. Durch Analyse nach dem Röntgenfluoreszenzverfahren konnte diese Vermutung bestätigt werden. Wie Bild 22 zeigt, liegt der Mangangehalt in den Streifen mit 0,15 bis 0,20 % erheblich über dem der angrenzenden Zonen und über dem Durchschnitt des kohlenstoffreichen Teiles. Das weist, ebenso wie die vielfach zu beobachtende Trennung durch Schlackeneinschlüsse darauf hin, daß die Streifen und die angrenzenden Zonen aus verschiedenen Luppen bestehen, die trotz der verhältnismäßig späten Herstellungszeit und Anwendung eines Verfahrens mit starker Reduktion ziemlich klein gewesen sind.

## Zusammenfassung

Die Arbeit sollte einen Überblick über die bei der Untersuchung alter Eisenfunde einzusetzenden Verfahren und die zweckmäßige Reihenfolge ihrer Anwendung geben. Sie zeigte, daß nicht, wie üblich, die chemische Analyse am Anfang stehen soll, sondern eine metallographische Untersuchung, die durch eine Härteprüfung ergänzt werden kann. Erst danach kann eine gezielte Analyse der einzelnen Zonen gemacht werden, bei der der metallische Teil und die eingeschlossenen Schlacken getrennt erfasst werden sollen. Wenn die zu untersuchenden Teile nicht zerstört werden dürfen, sind besondere metallographische Verfahren anzuwenden. In solchen Fällen kann auch mit Hilfe einer Durchstrahlung ein Einblick in den Aufbau des Stückes gewonnen werden. Bei der Entrostung ist Vorsicht geboten, weil aus Spuren im Rost manchmal noch das ursprüngliche Gefüge erkennbar ist, auch wenn das Stück völlig verrostet ist, sodaß nach einer Ablösung des Rostes nichts übrig bleiben würde.

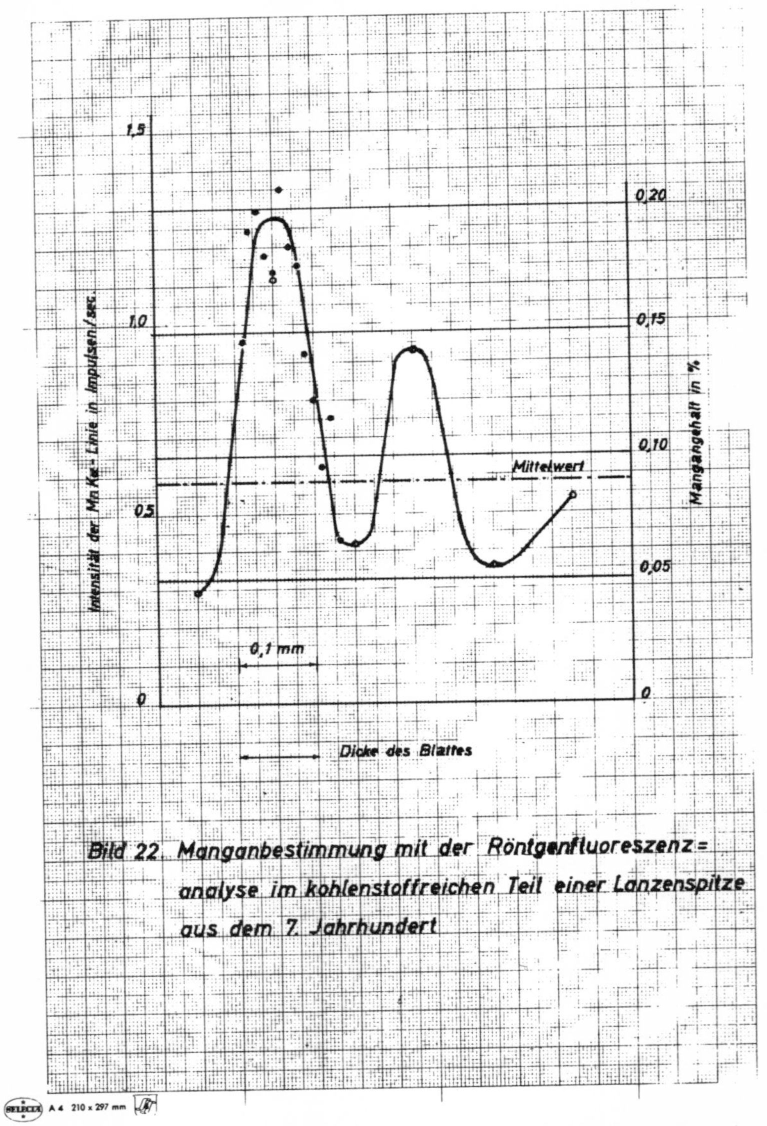


**Bild 22. Manganbestimmung mit der Röntgenfluoreszenz-analyse im kohlenstoffreichen Teil einer Langenspitze aus dem 7. Jahrhundert.**

# Research on an Iron Spearhead from the Etruscan Sanctuary of Fanum Voltumnae, Fourth to Third Centuries B.C.

By C. Panseri and M. Leoni
Istituto Sperimentale dei Metalli Leggeri
Novara, Italy

The spearhead, subject of the present research, was entrusted to us by the Superintendent of Antiquities of Etruria, Florence, and had been recovered at Montefiascone, in a tomb dating back to the fourth and third centuries B.C.

Montefiascone stands on a steep, tufaceous hill and has been identified by some students as the site of the famous sanctuary Fanum Voltumnae, the religious and political center of the Etruscan Confederation, where major heads of the *Dodecapoli* (confederation of twelve cities) met every year.

At the time of its recovery, the spearhead was covered with a thick, black patina, mamillary in appearance and easily removable by electrolytic treatment (Fig. 1). After removal of the incrustation, the spearhead was seen to be in a fairly good state of preservation. Corrosion had unevenly attacked the surface layers, showing up the most

**Fig. 1. Iron spearhead from Montefiascone, as recovered and after removal of the incrustation.**

hammered areas and leaving the central part almost unaltered. In the middle part of the tip, close to its fixture to the shaft hole, a slight ribbing may still be observed.

The more slender shaft hole, by means of which the tip was fixed to the shaft, had been almost destroyed by corrosion.

The head, classifiable as the laurel leaf type, presented the following characteristics:

— length of the head, from tip to beginning of widening of the blades, about 105 mm
— maximum width of the blade, about 40 mm
— thickness of the rib, in correspondance to the area of maximum width, about 8 mm
— weight of the head after removal of the incrustation, about 60 gm

## STRUCTURAL EXAMINATION

The sections on which the macrographic and micrographic examinations were carried out are shown in Fig 2. The macrographic examination was effected on sections 2 and 5, by etching the surfaces suitably polished with 5 percent Nital. The corresponding macrographs are illustrated in Figs. 3 and 4. As can be seen, the macrostructure of the sections has resulted in stripes, *i.e.* it consists of numerous steel layers of differing carbon percentages, welded together.

Since the corrosion caused by outside agents had acted sufficiently deeply and unevenly on the various zones of the head, and also because of the differing local chemical composition, it has not been possible to determine accurately the number of different layers with the varying amounts of carbon, which make up the head.

The micrographic examination of the Sections 1, 2, 3 and 4 of the head has, above all, enabled us to exclude the fact that the piece, subsequent to being forged, had undergone any complete or localized heat treatment in-

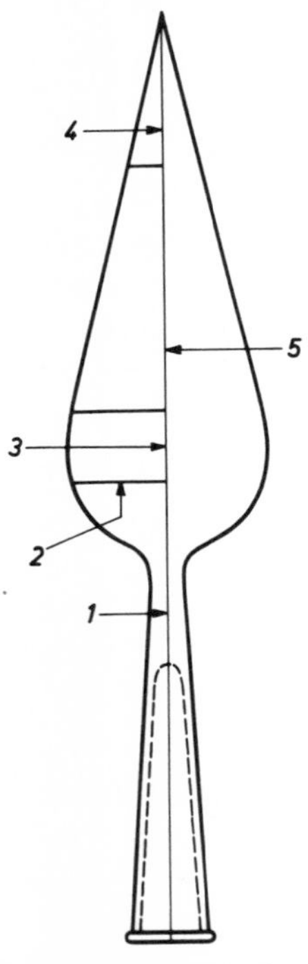


**Fig. 2. Design of the spearhead showing the sections submitted to examination.**

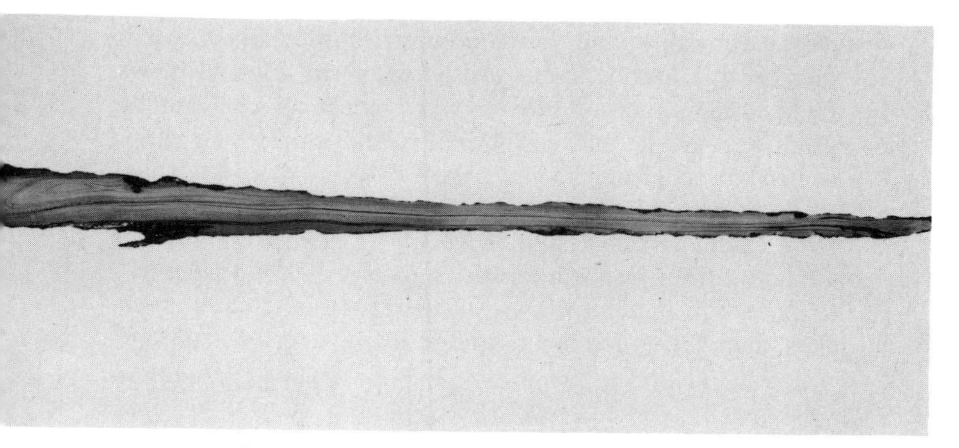

**Fig. 3. Macrostructure of section n. 5.**

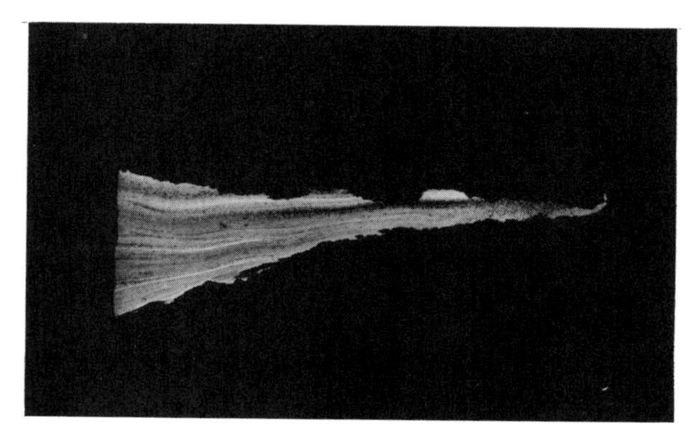

**Fig. 4. Macrostructure of section n. 2.**

tended to alter the mechanical characteristics (for example, carburizing and hardening, or hardening only) to increase the hardness and resistence of the edge to blows. In fact, the structural elements present in all zones of the piece were ferrite and perlite in a relationship which varied somewhat according to the zone.

As is seen in the micrograph made at limited enlargement (X50) in correspondence to the central part of the Section 2 (Fig. 5), the internal zone consisted of a layer at a high carbon content (0.4 – 0.5 percent) of greater thickness than the other layers and a mainly perlite structure.

**Fig. 5. Microstructure of central layers of the spearhead with high percentage of carbon.**

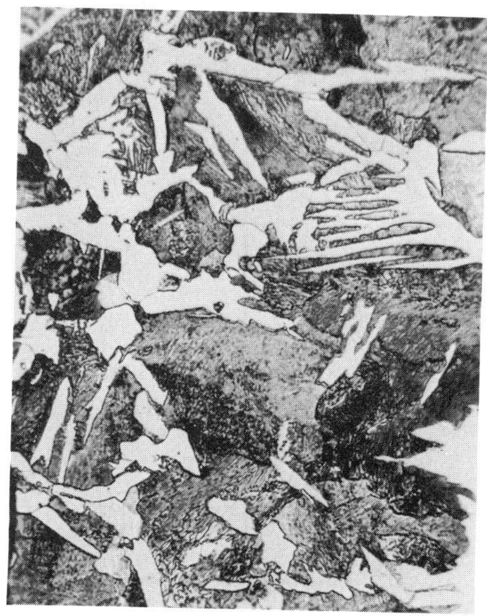

**Fig. 6. Microstructure of one layer with high percentage of carbon.**

Highly enlarged, the structure of this zone is illustrated in Fig. 6.

The Vickers microhardness, determined with a load of 1000 gm, resulted 200–215 Kg/mm.[2] The carbon percentage of the center layer decreased passing towards the adjacent layers because of the partial decarburization undergone during forging.

Proceeding towards the exterior, lateral to the center layer with a high carbon percentage, is noted a series of other layers of steels having more or less low carbon content. They are of ferrite or perlite-ferrite structure with carbon percentages of about 0.20–0.25 percent (Figs. 7, 8).

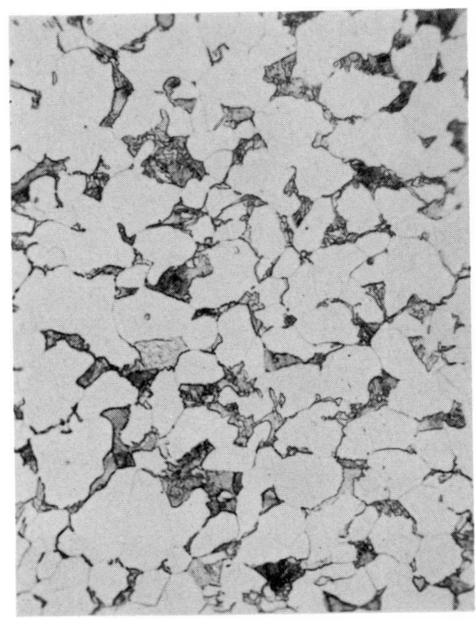

**Fig. 7. Microstructure of one layer with mean percentage of carbon.**

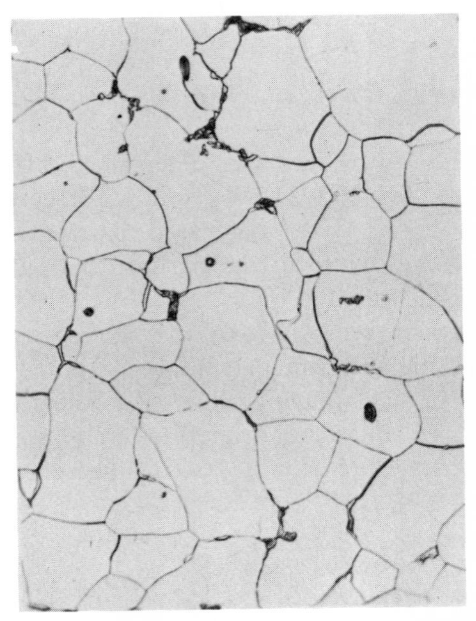

**Fig. 8. Microstructure of one layer with low percentage of carbon.**

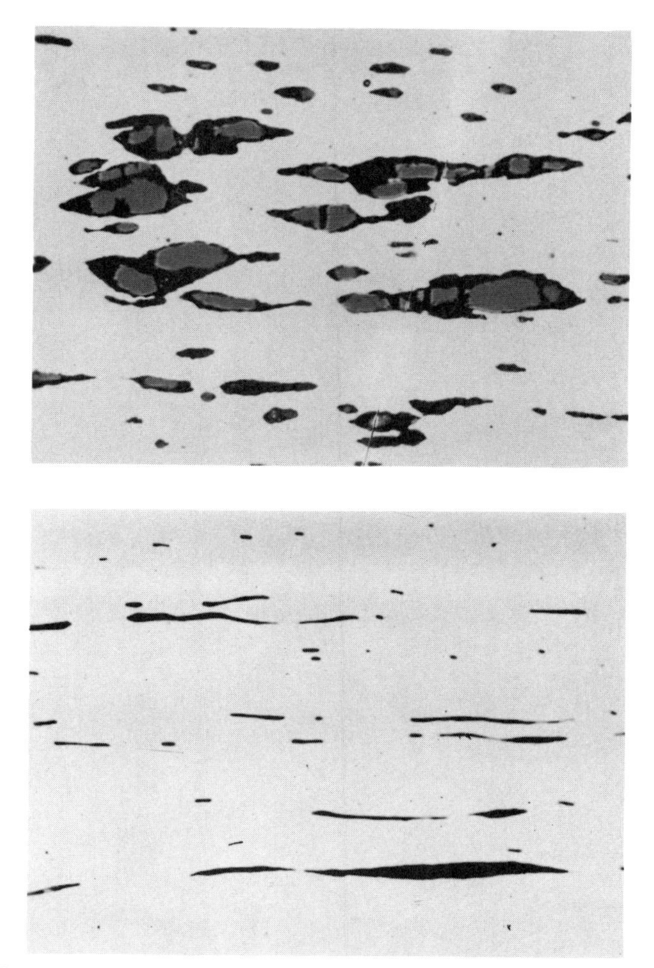

**Fig. 9. Microstructure of inclusions present in ferritic layers.**

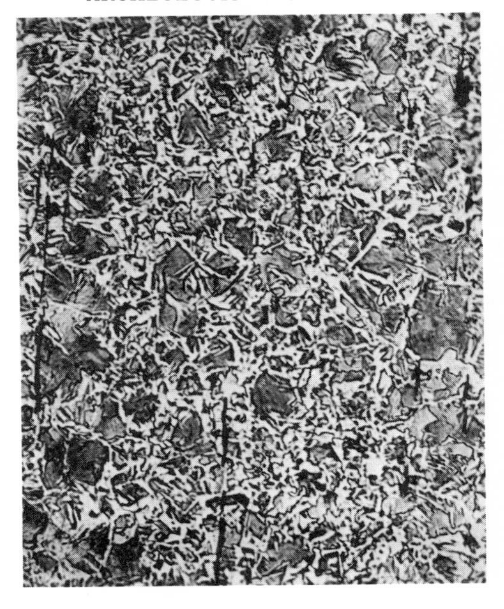

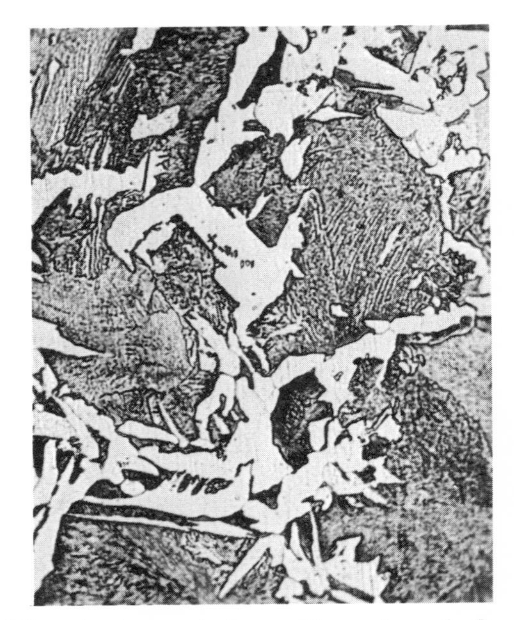

**Fig. 10. Microstructure of the cutting edge of the spearhead under examination.**

The layers with mainly ferrite structures showed a Vickers microhardness of 135–145 Kg/mm,[2] while the microhardness of those of perlite-ferrite structure resulted of 170–180 $Kg/mm^2$. The ferrite strips were particularly rich in slag inclusions of varying kinds, stretching in the direction of prevalent plastic deformation (Fig 9).

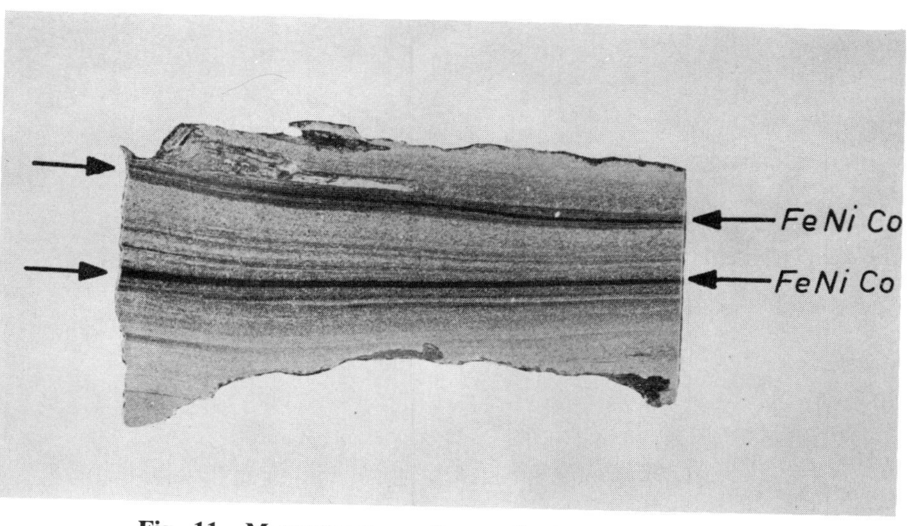


**Fig. 11. Macrostructure of a particular of section n. 3.**

By reason of the notable corrosion of the surface of the weapon, it was not possible to establish with certainty if the surface areas consisted of bands of hard or soft steel.

The other sections examined presented practically the same structure as the section described, *i.e.* they demonstrated that the differing layers of steel run through the whole head. The tip of the head had been obtained from the central area, noticeably carburized and of a high degree of hardness and perlite structure, as illustrated by Fig. 10.

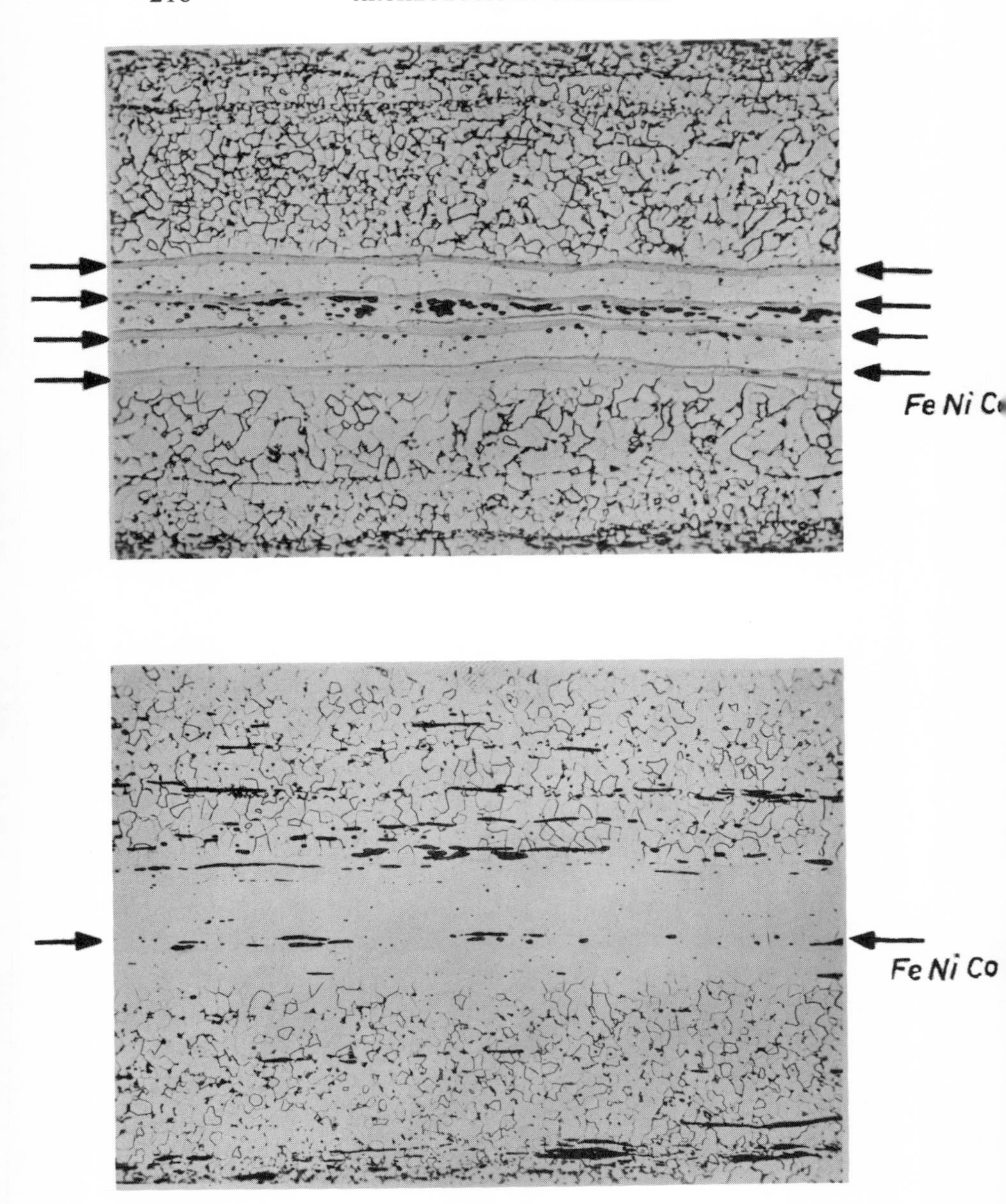


**Figs. 12 and 13. Microstructure of nickel-cobalt steel layers present.**

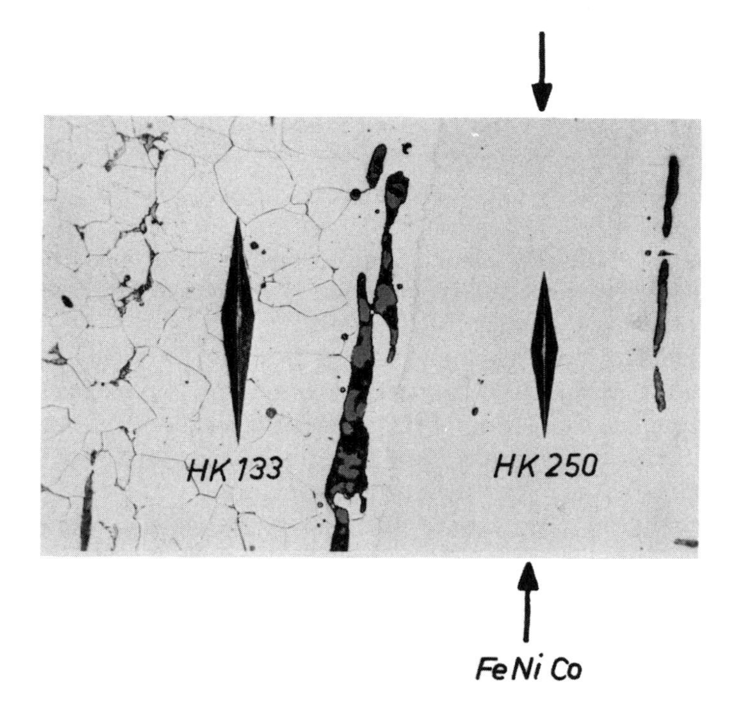


**Fig. 14. Knoop impressions obtained by a 100 g load upon ferritic and nickel-cobalt steel layers.**

As is seen in Fig. 11, lateral to the center zone, at about one-third of the thickness of the head, were present two fine layers of steel of different nature, being hardly etched by micrographic reagents and which thus appear more obscure in the photograph because more reflecting. These layers, the thickness of which varies almost in proportion to the variation of thickness of the head, were present in all sections examined and appeared on the surface at about 7–8mm from the cutting edge.

The micrographic examination, with a greater magnifying ratio, showed that one of these two layers consisted of four thinner layers divided by ferrite structure zones (Fig. 12). The Vickers microhardness test yielded a result between 240–250 $Kg/mm.^2$

In Fig. 13, two Knoop impressions are illustrated, effected with a load of 100 g on one of the above layers and the other on a ferrite structure zone.

To identify the chemical nature, a fragment of this particular type of steel has been taken away by means of preferential dissolution of the adjacent layers, and the spectrographic analysis has been made in comparison with an area of the head having normal structure.

It was thus ascertained that the layers consist of steel containing a certain percentage of nickel and a trace of cobalt (Fig. 14). The presence of these elements explains the low reactivity of these zones in comparison with the micrographic tests used for the common carbon steels. From the spectrographic test, the other zones of the head are shown to consist of very pure iron containing only minimum traces of manganese and copper. The high degree of iron purity is a particular characteristic of the iron handworks from archeological sources. This derives partly from the technique used by ancient peoples in the extraction of iron from its ores.[1]

The reasons for the presence in the head of these fine layers of steel of such special chemical composition are numerous and are not easy to explain. In fact, only a few hypotheses may be formulated; the alloy could have been formed by reduction of nickel and cobalt ores mixed with iron ores or, as is more probable, could be derived from a fragment of meteoric iron (siderite introduced into the pack from which the head was forged). The siderites consist normally of an iron-nickel alloy, with a somewhat variable percentage of nickel and traces of cobalt.

## EXAMINATION BY ELECTRON MICROBEAM PROBE

Being unable to carry out the quantitative analytical examination of the thin layers by normal analytical methods, because of their limited dimensions, we approached the *Institut de Recherches de la Sidérurgie* (IRSID) to carry out the analytical examination of the various bands of nickel-cobalt steel present in the head, by use of an electron microbeam probe.

For these tests, section 3 was sent, indicated in Fig. 9, the microstructure of which has already been illustrated in Fig. 11. As is noted, the electron microbeam analyser permits the rapid punctual elementary analysis of metal specimens, utilising X-ray spectrometry and using the specimen itself as anti-cathode. The accuracy of the analysis is about 1.1 percent, while the area analysed is about 2 $\mu$ in diameter.

The tests carried out have shown that the layers of ferrous-nickel-cobalt steel were, in turn, composed of various sub-layers of an alloy having an extremely variable nickel concentration. The concentration of cobalt always resulted at less than 0.1 percent.

The results of the analyses are shown in Table 1. It is observed that inside the layers, a scale of concentration indicated in the Table gives the highest values found at the center of the different sub-layers.

Apart from nickel and cobalt, researches were made also for chromium and manganese, but with negative results.

In our opinion, the presence of such a high percentage of nickel may be considered as an indisputable evidence of our second hypothesis about the siderite origin of the layers of ferrous-nickel-cobalt steel present in the spear-head examined.

Siderites, in fact, always contain a high percentage of nickel, the average value of which, according to specialized literature data [2] ranges from 4–33 percent.

The percentage of nickel may vary, however, from zone to zone of the siderite, because of the structural characteristics of these bodies.

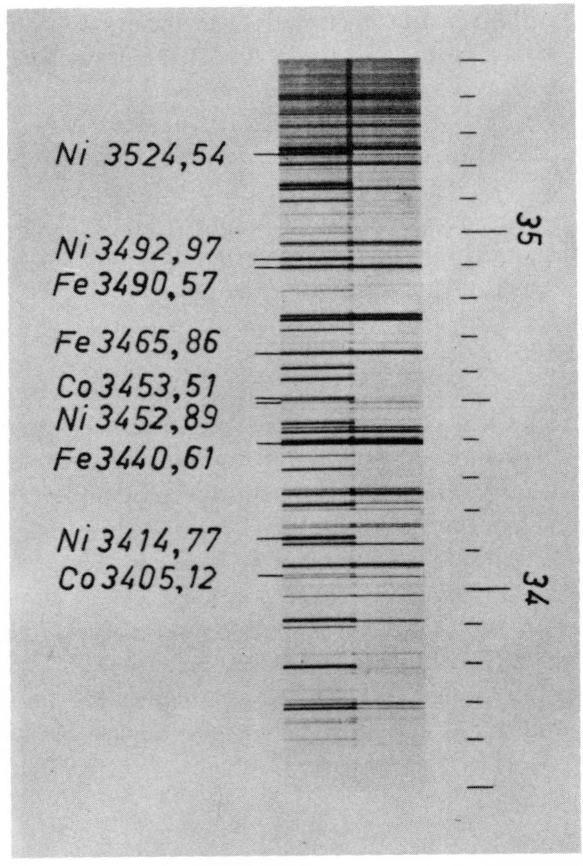


**Fig. 15. Detail of the spectral lines obtained in a zone of nickel-cobalt steel and in a zone of normal composition.**

The presence of meteoric iron in an object of archeological origin should not cause surprise; in fact numerous ancient hand-works of meteoric iron have been found which also belong to periods in which the extraction of iron from its ores had already begun.

The oldest iron objects, found in Egypt and belonging to the pre-dynastic age, are several tubular necklace beads of meteoric iron, containing an average of 7.5 percent of nickel.

## RADIOCRYSTALLOGRAPHIC EXAMINATION

To check the structure of the ferrous-nickel-cobalt layers by radiocrystallographic methods, too, an X-ray diffraction pattern by Debye-Scherrer method was recorded.

The radiocrystallographic specimen was thinned mechanically, and by chemical action, in correspondence with a layer of nickel-cobalt steel. From the layer reduced in this manner was cut a threadlike specimen. The experimental conditions were:

| | |
|---|---|
| Debye chamber | ∅ 57.3 mm |
| radiation | Cr K $\alpha$ (Vanadium filter) |
| current | 6 mA |
| tension | 35 Kv |
| time taken | 16 hours |

The debyegram obtained is reproduced in Fig. 16.

For each diffraction the corresponding interplanar spacing "d" has been calculated and the values obtained have been compared with the data of literature for $\alpha$ –Fe and with the values of "d" calculated for a c.f.c. lattice with $a_0 = 3.57$ Å, which belongs to the solid solution Fe–20% Ni.[3]

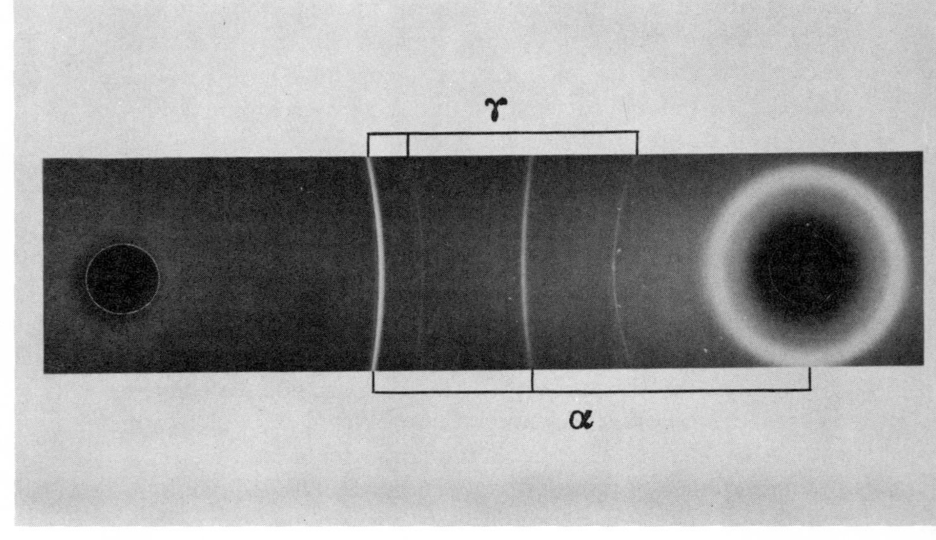


**Fig. 16. Debyegram obtained on a layer of nickel-cobalt steel present in the spearhead under examination.**

From comparison of the values reported in Table III, it is shown that in the debyegram of Fig. 16 there is present a very intense diffraction pattern of $\alpha$ Fe, and somewhat less intensely, also the pattern of $\gamma$ Fe–Ni. The lower intensity presented by the gamma phase may depend on its lower percentage or even simply from a reduced presence of this last phase on the surface of the radiocrystallographic specimen.

In fact, as illustrated by the micrograph in Fig. 12, the microscopic strands of nickel-cobalt steel alternate with ferrite bands of greater thickness from which it was not possible to separate them.

In addition, because of the wavelength of radiation used, the diffraction is active practically only on the outer layers of the specimen.

## CONCLUSIONS

From examination of the various sections of the spearhead, it can be concluded that it has been obtained by forging from a lump composed of a central layer of hard steel, having a carbon content of about 0.4–0.5 percent, to which had been joined laterally, by forge welding, two other lumps containing more sheets of steel at varying carbon contents.

That the spearhead has been made by welding of several steel foils of varying natures, intentionally positioned one over the other, is definitely proved by the presence of the two fine layers of meteoric steel found throughout the head. They are present, in fact, in all the sections examined. The probable schematic composition of the pack used for the forging of the weapon is illustrated in the diagram of Fig. 17.

From the lump thus composed has been forged the weapon evidencing the various layers by honing the spearhead to obtain the final shape required.

The top and cutting edges of the blades have been cut from the very hard perlitic central layer.

It was not possible to effect the examination of the shaft hole by means of which the spearhead was fixed to the shaft, as it is almost totally destroyed.

The results of the present investigations further confirm the conclusions previously reached, following our investigations on hand-works of Etruscan iron metallurgy.

The Etruscan blacksmiths were therefore in possession

of the so-called "welded Damascus" technique and knew, although empirically, the advantages of this technique. With the union in one pack of many steel layers at varying carbon contents, the hardness characteristics of a high carbon steel and the toughness of a soft steel are concentrated in one piece.

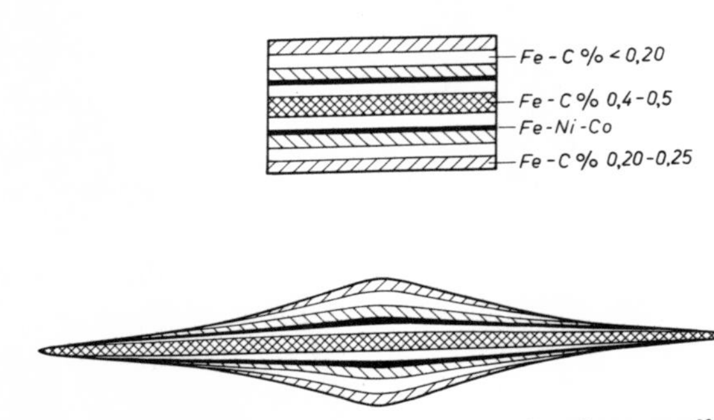


**Fig. 17. Schematic section of the spearhead according to the technological hypothesis deduced from the present metallographic research.**

In addition to being notably elastic and preserving well the sharpness of the cutting edges, blades made in this manner are not liable to break through brittleness as a result of blows received in combat, or to bend and become deformed like blades of softer materials, *e.g.* those typically used by the Celts.

This technique was to remain practically unchanged in Europe up to the end of the thirteenth century, *i.e.* until the coming of modern industrial metallurgy.[4]

Our preceding studies have also proved without doubt,

that already in the seventh century A.D., the curved swords of the archaic type (Kopis) were forged according to this technique, showing that the smiths who made them had long experience in the technology of iron. This period coincides with the beginning in Italy of the current production of weapons and iron hand-works. It may be considered that iron technology arrived in Etruria already considerably advanced and had probably been introduced by wandering smiths coming from the nearby Orient, attracted by the existence of rich ore deposits of siderite and cuprite in the maritime regions of Tuscany and Lazio. Populonia, with its port, was in fact, one of the main Mediterranean centers of the metal trade, first in copper and bronze, then in iron.[5]

Even if knowledge of the welded Damascus technique cannot be considered the prerogative of Etruscan metallurgy, it is certain that with this people it reached an extraordinarily high level for the times. It was known, even if in a less developed and rougher manner, by other populations, among them, those having a cultural "facies" of the Hallstatt type.

Different metallurgical cultural currents developed, originating from the same basic center of diffusion—from the mountains of the region presently called Armenia, rich in deposits of the metallic minerals that could be utilized in those times. In its expansion, this technique underwent considerable changes according to the living and cultural conditions of the various populations with whom it came into contact, and while in Etruria it was to reach a remarkable perfection, as has been seen, in the hands of other peoples it declined noticeably, almost to the point of extinction, as, for example, among the Gallic tribes [6] whose lack of efficiency with offensive weapons has been often mentioned by ancient historians.

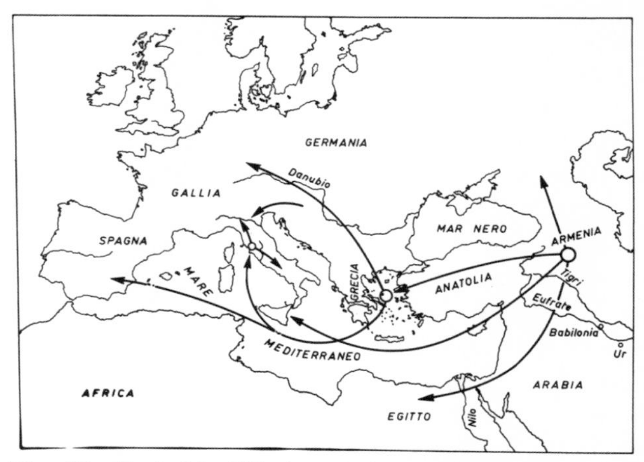


**Fig. 18. Probable diffusion of the metallurgical currents.**

A small axe found at Wietrzno-Bobrka in Poland, and belonging to the Hallstatt period, has also proved to have been made according to the welded Damascus technique. Moreover, in it has been found a thin layer of ferrous-nickel steel completely similar, even structurally, to those found by us in the spearhead referred to in this report. In this case, also, it has not been possible to ascertain the reasons for the presence of this particular type of natural steel in the weapon.[7] However, a suggestive, though not sufficiently substantiated hypothesis, could be drawn from the fact that meteoric iron, well known to the ancient peoples, was held to be of divine origin because it was recognized to come from heaven.

The Sumerian word for iron signified "metal of heaven" while, in Egyptian, iron was called "black copper of heaven." The introduction of even a small quantity of this material into a weapon, apart from its effective technical characteristics of hardness and toughness, could thus suggest in the mind of the primitive craftsman the acquisition of special magical powers and occult significance.

1. Panseri C., Leoni M., "Esame metallografico di alcune lame di spade galliche del II sec. a.C," *Metallurgia Ital.*, LI (genn. 1959), pp. 5–12.

2. Osmond F., Cartaud G., "Sur les fers météoritiques," *Rev. Métallurgie*, I (1094), pp. 69–79. Derge G., Kommel A. R., "The Structures of Meteoric Irons," *Amer. J. Sci.*, XXXIV (1937), p. 203. Gutenberg B., *Physics of the Earth*, "Internal Constitution of the Earth." McGraw-Hill Book Co. (New York 1939). Artini E., *I Minerali*, ed. U. Hoepli (Milano 1941), p. 316. Palache Ch., Berman H., Frondel C., *The System of Mineralogy*, Vol. I: *Elements, Sulfides Sulfosalts, Oxides* (New York 1946). Edwards A. B., *Textures of the Ore Minerals and Their Significance*, Australasian Inst. of Mining and Metallurgy (Melbourne 1954), p. 63. Ahrens L. H., Rankama K., Runcorn S. K., *Physics and Chemistry of the Earth* (London 1956), p. 241.

3. Pearson W. B., *A Handbook of Lattice Spacings and Structures of Metals and Alloys* (London 1958).

4. Perret J. I., *"L'art du coutelier,"* Paris (1771–1772).

5. Minto A., "L'antica industria mineraria in Etruria ed il porto di Populonia," *Studi Etruschi*, XXIII, serie II (1954), pp. 291–319.

6. Panseri, *op. cit.*

7. Piaskowski J., "A Socketed Axe from Wiertrzno-Bobrka in the Carpathians," *J. Iron & Inst.*, CXCIV, (mar. 1960), pp. 336–340.

TABLE I.

Chemical Composition of the Ferrous-Nickel-Cobalt Layers Analysed by Means of an Electron Microbeam Probe

| *Layers Analysed* | *Ni* % | *Iron* % | *Cobalt* |
|---|---|---|---|
| *I Layer* 2 sub-layer | 28.8 ± 1,8 | 70 ± 2 | ⩽ 0,1 |
| 4 sub-layer | 11.5 ± 1 | 88 ± 2 | ″ |
| 1 e 3 sub-layer | 0.8 ± 0,2 | 98 ± 2 | ″ |
| *II Layer* 2 , 4 , 6 sub-layer | 0.8 ± 0,2 | 97 ± 2 | ″ |
| 1 , 3 , 5 , 7 sub-layer | 11 ± 1 | 89 ± 2 | ″ |

TABLE II.

Atomic Percentage of Nickel Present in the Gamma Phases and Gamma at Different Temperatures

| *Temperature* °*C* | *Phase* α | | | | *Phase* γ | | | |
|---|---|---|---|---|---|---|---|---|
| 800 | 1.0 | 1.2 | 2.2 | — | 4.5 | 3.8 | 3.2 | — |
| 700 | 2.5 | 2.5 | 4 | — | 9 | 9.4 | 9.2 | — |
| 600 | 3.5 | 3.7 | 5.6 | 7.5 | 14 | 17.3 | 18.5 | — |
| 550 | 4.4 | 4.4 | 6.3 | 8.8 | 18.5 | 22.2 | 24 | — |
| 500 | 5.0 | 5.0 | 7.1 | 10.1 | 26.5 | 27.5 | 29.5 | 28.2 |
| 450 | 5.7 | 5.7 | 7.8 | 11.5 | 34 | 34.3 | 33–35 | 30.7 |
| 400 | 6.5 | 6.3 | 8.5 | 13.2 | 41.5 | 41.5 | — | 32.7 |
| 350 | 5.8 | 6.9 | — | 15.5 | 49 | 48.5 | — | 34.4 |
| 300 | 4.8 | 7.5 | — | — | 56.5 | 56 | — | — |
| *Reference* | (9) | (10) | (11) | (12) | (9) | (10) | (11) | (12) |

TABLE III.

Intensity and Interplanar Spacings of the Debyegram Lines of Fig. 5

| *Experimental Values* | | $\alpha - Fe$ | | $\gamma - Fe - Ni$ $a_0 = 3.57$ |
|---|---|---|---|---|
| I | d | I | d | d |
| *dd* | 2.053 | | | 2.06 |
| *f* | 2.024 | 100 | 2.01 | |
| *d* | 1.789 | | | 1.79 |
| *sf* | 1.433 | 15 | 1.431 | |
| *d* | 1.265 | | | 1.26 |
| *ff* | 1.170 | 38 | 1.168 | |

dd = very weak
d = weak
sf = semi-strong
f = strong
ff = very strong

# Investigations on Ancient Pottery from Maski

By Dr. S. Paramasivan
Madras, India

B. K. Thapar has propounded the existence of three distinct cultural levels at Maski * which corresponds to three different periods:

Period I—Chalcolithic culture, early first millennium B.C. to 400 B.C.
Period II—Megalithic culture 200 B.C. to first century A.D.
Period III—Early historical culture, middle of the first century A.D. to third century A.D.

Among other things, pottery pieces have been unearthed from the different cultural levels. Since pottery constitutes the very alphabet of archaeology, typical samples of them were taken up for study and investigation with a view to working out their physical and chemical characteristics and through them to effect, if possible, an interpretation and a correlation. Since no work of this type has, so far, been attempted anywhere else, an investigation of this nature is all the more necessary. The real inspiration for this work came through a formal study of the remarkable types of Cretan and Mycenaean pottery in Athens and elsewhere.

This study has its limitations. Study of pottery (as also of other archaeological materials) is essentially a statistical subject. Frequencies of occurrence of the different types must be worked out, and samples for investigation must be taken from the more commonly occurring ones. With any other sort of approach, study of pottery loses much of its scientific value.

In connection with these investigations, the following sherds, corresponding to the three periods referred to above, were taken up for study:

Period I—

Painted ware: 2 sherds, M1, M2.
Burnished grey ware: 4 sherds, M3, M4, M5, M6.
Pinkish buff ware: 4 sherds, M7, M8, M9, M10.

Period II—

Megalithic black-and-red ware from habitation site: 1 sherd, M11.
Megalithic black-and-red ware from burial site: 4 sherds, M12, M13, M14, M15.

Period III—

Russet-coated painted ware: 3 sherds, M16, M17, M18.

Thus 18 sherds have been taken up for study and investigation. In the following discussions, the sherds will be referred to as M1, M2, M3 etc. M1 to M10 belong to Period I. M11 to M15 belong to Period II. M16 to M18 belong to Period III.

## Colours of Sherds

One of the important archeological characteristics of pottery is colour. Colours are normally specified through their visual impression. But this method is subjective. It may not mean the same thing to all. It was, therefore, felt necessary to specify the colours of the sherds according to some standard system, so that they may mean the same thing to all archeologists wherever they might be, without any ambiguity. Two systems have been proposed to this purpose:

(1) The British Standards Institute have issued a series of standard painted cards to illustrate the colours of different paints. All the different gradations of colour are represented. Each card has an index number. By comparing the colours of the potsherds with the colours on these cards, the former may be specified accurately in terms of the standard colour cards and the corresponding indices associated with them.

(2) The colours of potsherds may be compared with the colours of optical spectra and specified in relation to them, that is, in terms of the corresponding wave lengths in angstrom units.

Colour Values According to the British Standard Specification

| | |
|---|---|
| *M1* | Terra Cotta No. 444 |
| *M2* | Terra Cotta No. 444 |
| *M3* | Middle Graphite No. 671 |
| *M4* | Middle Graphite No. 671 |
| *M5* | Middle Graphite No. 671 |
| *M6* | Middle Graphite No. 671 |
| *M7* | Deep Buff No. 360 |
| *M8* | Deep Buff No. 360 |
| *M9* | Light Brown No. 410 |
| *M10* | Deep Buff No. 360 |
| *M11* | Dark Blue Grey No. 695 & Light Brown No. 410 |
| *M12* | Dark Blue Grey No. 695 & Light Brown No. 410 |
| *M13* | Dark Blue Grey No. 695 & Light Brown No. 410 |
| *M14* | Dark Blue Grey No. 695 & Light Brown No. 410 |
| *M15* | Dark Blue Grey No. 695 & Light Brown No. 410 |
| *M16* | Red Oxide No. 446 |
| *M17* | Red Oxide No. 446 |
| *M18* | Imperial Brown No. 415 |

If intervening colour cards were obtainable showing subtler variations in colours, the colours of sherds could be specified to a finer degree of accuracy. With standard colour cards, which may be purchased easily, and the index numbers which may be easily fixed to the fabrics, one can visualise the exact colour of the sherds.

Colour Values According to the Optical Spectrum in the Visible Region

| | |
|---|---|
| *M1* | 6000–6249 A.U. |
| *M2* | 6000–6249 A.U. |
| *M3* | 7313.01–7664.91 A.U. |
| *M4* | 7313.01–7664.91 A.U. |
| *M5* | 7664.91–7800.23 A.U. |
| *M6* | 7513.01–7664.91 A.U. |
| *M7* | 6249.93–6438.48 A.U. |
| *M8* | 6249.93–6438.48 A.U. |
| *M9* | 6249.93–6438.48 A.U. |
| *M10* | 6249.93–6438.48 A.U. |
| *M11* | 3961.53–4000 A.U. |
| *M12* | 3961.53–4000 A.U. |
| *M13* | 3961.53–4000 A.U. |
| *M14* | 3961.53–4000 A.U. |
| *M15* | 3961.53–4000 A.U. |
| *M16* | 6500–6707.85 A.U. |
| *M17* | 6500–6707.85 A.U. |
| *M18* | 6707.85–6911.3 A.U. |

Of the two methods, the former is easier of application and quite sufficient for all archaeological purposes.

Some suggestions have been made that colour transparencies of typical potsherds may be used to compare the colours of unknown pottery with. This method has its limitations. Transparencies cannot give faithful reproductions of colour in all cases. They cannot bring out that latitude of colour range which can be through these standard cards. The transparencies suffer in quality through time.

## THICKNESS OF FABRIC AND SLIP

The thickness of the fabric depends on the nature of the ware. But given a sample, the ratio of the maximum to minimum thickness, especially in the case of a small sherd which is what is usually sent for examination, is of special significance. Wherever uniform or nearly uniform thickness is maintained, it is an indication that the artisan has very skilfully worked the fabric. In the absence of uniformity of thickness, one may presume that the workmanship is relatively poor or was entrusted to amateur hands.

The thicknesses were measured in the case of all the potsherds.

| | *Fabric* | | |
|---|---|---|---|
| *Specimen No.* | *Maximum* | *Minimum* | *Slip* |
| *M1* | 0.398 mm | 0.354 mm | 0.005 mm |
| *M2* | 0.491 | 0.334 | 0.004 |
| *M3* | 0.579 | 0.452 | 0.003 |
| *M4* | 1.061 | 0.948 | 0.003 |
| *M5* | 1.480 | 0.981 | 0.002 |
| *M6* | 1.182 | 0.888 | 0.002 |
| *M7* | 0.965 | 0.717 | 0.003 |
| *M8* | 0.916 | 0.689 | 0.003 |
| *M9* | 1.142 | 0.698 | 0.003 |
| *M10* | 0.766 | 0.596 | 0.002 |
| *M11* | 0.432 | 0.331 | 0.003 |
| *M12* | 0.399 | 0.347 | 0.002 |
| *M13* | 0.327 | 0.313 | 0.002 |
| *M14* | 0.369 | 0.333 | 0.002 |
| *M15* | 0.446 | 0.441 | 0.001 |
| *M16* | 0.519 | 0.438 | 0.007 |
| *M17* | 0.639 | 0.617 | 0.005 |
| *M18* | 0.922 | 0.918 | 0.004 |

Here M1 and M4 of Period I, M12 to 15 of Period II, and M17 and M18 of Period III have almost uniform thickness, thereby showing manipulative skill in workmanship. This is evident from the fact that there is not much difference between the maximum and minimum values.

The fabrics of Period I may be divided into three groups. The first group consists of M1 and M2 with the thinnest sections. The next group, which consists of M7 and M8 have sections of intermediate thickness. The thickest sections occur in M5, M6 and M9, which belong to the third group. Period II has thinner fabrics. The thickness of this fabric is comparable with that of group I of Period I. Period III has fabrics which are somewhat intermediate in thickness between some of the wares of Period I, namely, M4, M5, M6 etc. and M11 to M15 of Period II.

In a similar manner, the thickness of the slip ranges between 0.002 to 0.003 m.m. But M1 and M2 of Period I and M16 and M18 of Period III have relatively abnormal thicknesses. In such cases, the slip has probably not been sufficiently diluted with water. It may be that the preparation of the slip was not stirred up and used immediately, without allowing time for the clay particles to settle down as much as possible. But such abnormal values are not common, only exceptional.

The binding between the body and the slip is quite strong, thereby showing that the slip is probably of the same material as the body.

It may be mentioned here that uniformity of thickness is also an indication that the fabric has been turned on the wheel or, if the ware was hand-made, the potter had an extremely dexterous hand.

## HARDNESS, SPECIFIC GRAVITY AND POROSITY

The hardness, specific gravity and porosity of the sherds are tabulated below:

| *Specimen* | *Hardness* | *Specific Gravity* | *Porosity* |
|---|---|---|---|
| *M1* | 2.5 – 3 | 2.810 | 19.62 |
| *M2* | 2.3 – 3 | 2.921 | 18.47 |
| *M3* | 2.5 – 3 | 2.892 | 23.39 |
| *M4* | 2.5 – 3 | 2.941 | 18.92 |
| *M5* | 2.5 – 3 | 2.901 | 17.39 |
| *M6* | 2.5 – 3 | 2.896 | 22.12 |
| *M7* | 2.5 – 3 | 2.940 | 24.76 |
| *M8* | 2.5 – 3 | 2.879 | 20.04 |
| *M9* | 2.5 – 3 | 2.896 | 18.79 |
| *M10* | 2.5 – 3 | 2.921 | 19.23 |
| *M11* | 3.0 – 3.5 | 2.623 | 14.67 |
| *M12* | 3.5 – 4 | 2.686 | 19.02 |
| *M13* | 3.5 – 4 | 2.707 | 16.29 |
| *M14* | 3.5 – 4 | 2.628 | 15.77 |
| *M15* | 3.5 – 4 | 2.876 | 15.64 |
| *M16* | 3.0 – 3.5 | 2.742 | 14.21 |
| *M17* | 3.0 – 3.5 | 2.723 | 17.43 |
| *M18* | 3.0 – 3.5 | 2.719 | 18.02 |

The hardness of the specimens varies from 2.5 to 3.0 in the case of samples M1 to M10. With specimens of Periods II and III it varies from 3 to 4. This probably means that in the case of the latter, hard minerals are more predominant.

Ceramics may be roughly divided into pottery proper and porcelain. Each of them may be divided into hard and soft, probably with intermediate stages. The differences are attributed to differences in the mineral composition of the clay and the degree of heat to which it has been subjected in the course of firing. The result is that there is a difference in hardness of surface and in the power of resisting the action of fire. With the exception of M11, the pottery pieces of Period II are the hardest of the lot. Those of Period I are the least hard. The pottery pieces of Period III occupy an intermediate position.

The specific gravity of the sherds of Period I varies from

2.810 to 2.941. The specific gravity of clay varies from 2.623 to 2.876 for Period II. The values for Period III are almost the same. This indicates that probably almost the same variety of clay, or clay from almost the same locality has been used throughout.

In many cases, the porosity varies within narrow limits, namely from 17 to 20 for Period I. But M3, M6 and M7 exhibit somewhat abnormal values. For Period II, the value ranges between 14.67 for M11 and 19.02 for M12. With the exception of M12, the other sherds of Period II have values which lie within the range of the values for Period I. The value varies from 14.21 to 18.02 for Period III, the value for M16 being somewhat abnormal as compared with the values for M17 and M18.

The samples which exhibit abnormal values for porosity are as follows: M3 (22.39), M6 (22.12), M7 (24.76), M11 (14.67), M14 (15.27), M15 (15.64), M16 (14.21).

Porosity depends on the size of the particles. The more coarse the particles are, the larger is the space between them and higher the porosity. In the list showing the abnormal values, the first three belonging to Period I have coarse materials in the fabric. The next three belong to Period II and the last one to Period III. All the other samples have relatively finer particles.

## THERMAL PROPERTIES

*Variation of specific gravity with temperature.* Fragments of each of the samples were heated to 400°, 500°, 600°, 700°, 800°, 900°, 1000°, 1100° and 1250° C. respectively, and the specific gravity and porosity corresponding to the particular temperature determined.

There is little or no change in specific gravity with temperature, as illustrated in the following typical cases:

| Specimen No. | 400° | 500° | 600° | 700° | 800° | 900° | 1000° | 1150° | 1250° |
|---|---|---|---|---|---|---|---|---|---|
| *M1* | 2.943 | 2.927 | 2.921 | 2.901 | 2.897 | 2.884 | 2.876 | 2.989 | 3.034 |
| *M2* | 2.898 | 2.891 | 2.887 | 2.884 | 2.880 | 2.889 | 2.894 | 2.942 | 2.972 |
| *M3* | 2.904 | 2.899 | 2.890 | 2.884 | 2.879 | 2.871 | 2.882 | 2.897 | 2.912 |

It is evident that, in the region of temperature up to which the sherds are heated, there is not much of a chemical change in the material, with the result that the specific gravity remains almost unaltered. There is one instance where the value is somewhat abnormal. It is with respect to M1 of Period I, the maximum occurring at 1250° C. It shows that there has been a contraction of the material on heating, with a consequent rise in specific gravity.

*Variation of porosity with temperature.* The porosity increase at first, reaches a maximum and then falls thus:

| *Specimen No.* | 400° | 900° | 1000° | 1100° | 1250° |
|---|---|---|---|---|---|
| *M1* | 23.57 | 30.91 | 31.12 | 22.37 | 16.44 |
| *M2* | 22.93 | 32.21 | 32.49 | 25.47 | 18.48 |
| *M3* | 25.09 | 31.01 | 31.87 | 24.77 | 18.54 |
| *M4* | 20.13 | 29.94 | 30.09 | 25.79 | 16.24 |
| *M5* | 19.21 | 28.91 | 29.56 | 20.04 | 15.43 |
| *M6* | 23.47 | 32.24 | 32.79 | 26.37 | 16.31 |
| *M7* | 26.39 | 35.77 | 26.92 | 19.43 | 14.37 |
| *M8* | 24.17 | 29.64 | 27.32 | 20.31 | 16.34 |
| *M9* | 22.30 | 31.10 | 25.49 | 19.92 | 15.96 |
| *M10* | 23.60 | 30.14 | 26.19 | 23.96 | 15.52 |
| *M11* | 16.79 | 20.79 | 18.41 | 15.69 | 12.45 |
| *M12* | 16.43 | 21.96 | 20.43 | 16.49 | 11.79 |
| *M13* | 16.96 | 21.29 | 18.64 | 16.32 | 12.39 |
| *M14* | 16.31 | 22.43 | 16.39 | 13.96 | 12.02 |
| *M15* | 17.46 | 27.43 | 20.71 | 16.39 | 11.48 |
| *M16* | 16.29 | 29.12 | 29.34 | 20.71 | 12.32 |
| *M17* | 18.45 | 27.81 | 28.29 | 20.12 | 13.49 |
| *M18* | 19.06 | 30.21 | 30.45 | 24.43 | 13.21 |

The porosity of the samples M1 to M6 of Period I and of M16 to M18 of Period III reaches a maximum at 1000° C. The values for the samples M7 to M10 of Period I and M11 to M15 of Period II reach a maximum at 900° C. The results are significant. At the higher temperatures the fabric softens, expands and melts in all probability and the pores are somewhat blocked. It might also mean that there is a virtual contraction of the fabric and a closing of the pores at the temperature corresponding to the minima in the values for porosity.

## OPTICAL PROPERTIES

Micro-sections of the different samples of pottery were prepared and their optical properties studied under a polarising microscope. These studies give information regarding (a) composition of the clay, (b) character of the minerals and (c) grain size of the minerals commonly occurring.

(a) *Composition of the clay.* In addition to quartz and muscovite, the samples contain the following minerals:

| | |
|---|---|
| *M1* | Altered biotite, sericite, chlorite. |
| *M2* | Altered biotite, sericite, chlorite. |
| *M3* | Altered biotite (kaolinite and montmorillinite). |
| *M4* | Biotite, orthoclase, kaolinite, montmorillinite. |
| *M5* | Orthoclase, kaolinite, montmorillinite. |
| *M6* | Kaolinite, montmorillinite. |
| *M7* | Biotite, kaolinite, montmorillinite. |
| *M8* | Kaolinite, montmorillinite, zircon and rutile. |
| *M9* | Biotite, kaolinite, montmorillinite, felspar, zircon. |
| *M10* | Biotite, kaolinite, montmorillinite, felspar, zircon. |
| *M11* | Felspar, hornblende, carbonaceous matter. |
| *M12* | Hornblende, carbonaceous matter. |
| *M13* | Biotite, kaolinite, carbonaceous matter. |
| *M14* | Biotite, hornblende, carbonaceous matter. |
| *M15* | Carbonaceous matter. |
| *M16* | Biotite, hornblende, kaolinite, carbonaceous matter. |
| *M17* | Biotite, hornblende, kaolinite, carbonaceous matter. |
| *M18* | Biotite, hornblende, kaolinite, carbonaceous matter. |

The specimens M11 to M15 belonging to Period II and

M16 to M18 of Period III contain some carbonaceous matter. Probably, it was not completely oxidised in the course of firing in the kiln. It is equally probable that the atmosphere in the kiln was more or less a reducing one. The presence of carbonaceous matter is interesting. Clays of fluviatile origin form the alluvium found along the river courses and on the flood plains. They vary greatly in composition and usually carry a large quantity of organic and especially vegetable matter. Many delta and lacustrine deposits are of an essentially similar nature.

Again, granite rocks with felspar as matrix disintegrate and leave behind the clay mixed with more or less resistant varieties of mica, quartz and other minerals which originally formed the granite rock. The more important agents which facilitate the decomposition and disintegration of the alumino-silicates are volcanic gases, water draining from peat bogs and coal beds, and spring or rain water containing carbon dioxide in solution. It is the third factor which is responsible here for the disintegration of the granite rock and the formation of clay. It is significant that some of the samples contain large grains of mica.

(b) *Character of minerals.* In all the samples, the minerals were angular to sub-angular. There were no rounded grains at all. In the absence of rounded grains, there is good interlocking of the minerals composing the clay, and this contributes towards the firmness and strength of the body. The absence of rounded grains is also an indication that the clay was not water-borne, though it may be lacustrine.

(c) *Grain size of commonly occurring minerals.* In each of the samples, the grain sizes of the commonly occurring minerals were measured, and tabulated as below:

| *Samples* | *Quartz* | *Muscovite* | *Other Minerals* |
|---|---|---|---|
| *M1* | .050 – .005 mm | .010 – .003 mm | .004 – .001 mm |
| *M2* | .050 – .005 | .020 – .005 | .004 – .001 |
| *M3* | .250 – .003 | .005 – .001 | .010 – .001 |
| *M4* | .050 – .005 | .005 – .001 | .010 – .001 |
| *M5* | .025 – .005 | .015 – .001 | .010 – .001 |
| *M6* | .030 – .005 | .020 – .003 | .015 – .001 |
| *M7* | .050 – .005 | .010 – .001 | .020 – .001 |
| *M8* | .050 – .005 | .020 – .001 | .015 – .001 |
| *M9* | .060 – .005 | .005 – .001 | .005 – .001 |
| *M10* | .080 – .005 | .010 – .002 | .025 – .001 |
| *M11* | .020 – .002 | .010 – .002 | .015 – .001 |
| *M12* | .020 – .002 | .020 – .002 | .020 – .001 |
| *M13* | .030 – .005 | .010 – .002 | .005 – .001 |
| *M14* | .020 – .002 | .010 – .002 | .005 – .001 |
| *M15* | .020 – .002 | .010 – .002 | .015 – .001 |
| *M16* | .050 – .003 | .010 – .002 | .008 – .001 |
| *M17* | .030 – .012 | .005 – .001 | .005 – .001 |
| *M18* | .030 – .002 | .005 – .001 | .005 – .001 |

The fabrics of M1, M2, M4 and M7 to M9 of Period I, and of M16 of Period III have quartz grains which are almost similar in size. The fabrics of M5 and M6 of Period I, of M11 to M15 of Period II and of M17 and M18 of Period III have quartz grains which are somewhat larger. The largest quartz grains occur in M3 of Period I.

The fabrics of M3, M4 and M9 of Period I have the smallest grains of muscovite. Next in order come the fabrics of M1, M7 and M10 of Period I, of M11, M13 to M15 of Period II and of M16 of Period III which have slightly larger particles. Still larger particles occur in the fabrics of M2, M5, M6 and M8 of Period I and M12 of Period II.

So far as other minerals are concerned, the smallest grains occur in M1, M2, M9 of Period I, M13 and M14 of Period II and M17 and M18 of Period III. Somewhat larger grains occur in M3 to M5 of Period I. Still larger grains occur in M6 to M8 and M10 of Period I and in M11, M12 and M15 of Period II.

Fabrics showing the ascending order of grain sizes are as follows:

| | |
|---|---|
| *Quartz* | M1, M2, M4, M7, M8, M9, M16 |
| | M11, M12, M13, M14, M15, M17, M18 |
| | M3 (largest) |
| *Muscovite* | M3, M4, M9 |
| | M1, M7, M10, M11, M13, M14, M15, M16 |
| | M2, M5, M6, M8, M12 |
| *Other Minerals* | M1, M2, M9, M13, M14, M17, M18 |
| | M3, M4, M5 |
| | M6, M7, M8, M10, M11, M12, M15 |

Taking the grain sizes as such in all the fabrics, the megalithic ware corresponding to M11 to M15 of Period I are of smallest grain and hence of finer fabric than the ware of Period I or Period III.

## CHEMICAL COMPOSITION

Pure clays exist only in theory. Clay always has impurities. The ingredients of typical clay are as follows:

| | |
|---|---|
| *Silica* | 60% |
| *Alumina* | 30 |
| *Iron* | 7 |
| *Lime* | 2 |

The chemical composition of all the potsherds has been worked out and tabulated below, and may be compared with the composition of almost pure clay given above. The percentage of silica occurring in the various potsherds is quite typical of the composition of pure clay. Iron and alumina occur, however, to much smaller extent.

| *Samples* | *Loss on Ignition* | $SiO_2$ | $Al_2O_3$ | *FeO* | $Fe_2O_3$ | *CaO* | *MgO* | *Alkalies* |
|---|---|---|---|---|---|---|---|---|
| *M1* | 11.64% | 55.53% | 17.94% | 5.36% | 2.74% | 3.39% | 1.76% | 1.56% |
| *M2* | 11.72 | 55.32 | 17.47 | 5.89 | 2.52 | 3.96 | 1.87 | 1.64 |
| *M3* | 9.87 | 57.19 | 17.09 | 4.92 | 2.26 | 4.24 | 2.46 | 1.87 |
| *M4* | 10.96 | 57.06 | 17.25 | 4.86 | 2.32 | 3.96 | 2.59 | 1.84 |
| *M5* | 10.30 | 57.02 | 16.92 | 4.95 | 2.07 | 4.43 | 2.42 | 1.82 |
| *M6* | 9.76 | 56.02 | 17.91 | 4.39 | 2.53 | 4.66 | 2.70 | 1.93 |
| *M7* | 10.29 | 55.87 | 17.35 | 4.61 | 2.45 | 4.67 | 2.66 | 1.97 |
| *M8* | 10.57 | 55.94 | 18.06 | 4.32 | 2.39 | 4.17 | 2.70 | 1.90 |
| *M9* | 10.19 | 55.76 | 17.42 | 4.74 | 2.66 | 4.59 | 2.59 | 1.91 |
| *M10* | 10.19 | 54.82 | 16.85 | 4.36 | 2.49 | 4.62 | 2.79 | 1.73 |
| *M11* | 11.85 | 54.09 | 16.97 | 4.91 | 3.02 | 4.69 | 2.46 | 1.87 |
| *M12* | 11.93 | 54.39 | 16.73 | 4.64 | 3.11 | 4.74 | 2.39 | 1.79 |
| *M13* | 11.92 | 54.94 | 17.63 | 5.72 | 2.60 | 3.67 | 1.70 | 1.39 |
| *M14* | 12.07 | 54.31 | 16.92 | 4.51 | 3.39 | 4.59 | 2.45 | 1.82 |
| *M15* | 12.02 | 54.58 | 16.87 | 4.62 | 2.92 | 4.93 | 2.17 | 1.84 |
| *M16* | 10.25 | 55.99 | 17.64 | 5.82 | 3.25 | 3.76 | 1.50 | 1.73 |
| *M17* | 10.79 | 55.81 | 17.57 | 5.96 | 3.02 | 3.55 | 1.42 | 1.69 |
| *M18* | 10.32 | 55.43 | 17.79 | 6.09 | 3.41 | 3.59 | 1.62 | 1.74 |

The clays used during all three periods had a large proportion of alumina. The proportion of iron is just half that of alumina. The more aluminous a clay is, the more plastic it is, and the easier the potter will find it to work with. The samples also contain a significant proportion of magnesia. With a larger proportion of magnesia, the clay would belong to the group of "burnt clays," which are the more abundant varieties of clay. The samples also contain fluxes like the compounds of iron and magnesium, which promote fritting or incipient marginal fusion of the particles on being burnt, thus binding the particles firmly.

The clay of M6 and M10 of Period I is a better aluminous type of clay. M3 to M5 and M7 to M9 of Period I and M11, M12 and M15 of Period II constitute the less aluminous type of clay. The still less aluminous type occur in M1 and M2 of Period I, in M13 and M14 of Period II, and in M17 of Period III. The least aluminous type of clay occurs in M18 of Period III.

The loss on ignition is minimum in M3 and M6 of Period I and maximum in M10 of Period I and M15 of Period II. Other samples have intermediate values. The loss on ignition may be due to carbonaceous matter or to free and combined water or to all three causes.

The proportion of silica is almost constant in the various samples and varies between 54.09 and 57.06 percent. Ferrous oxide (FeO) varies in value between 4.39 and 6.09, most of the specimens having intermediate values. Ferric oxide ($Fe_2O_3$) varies in value between 2.07 and 3.39, most of the samples having intermediate values. The samples of Periods II and III have more uniform values.

Lime (CaO) varies in value between 3.39 and 4.66. The values are more uniform for samples of Period II. Magnesia (MgO) varies in value from 1.42 to 2.79 in the various samples. But the samples of Period II have more uniform values. The proportion of alkalies in the different specimens does not vary very much.

In the various samples, the proportion of major constituents does not vary very much and hence it seems that the clay for them must have been taken from almost the same spot.

The presence of Ferric oxide ($Fe_2O_3$) and lime ($CaO$) is of special significance. A clay which contains 1 to 3 percent of $Fe_2O_3$ is buff-coloured after firing. It becomes red with 4 to 5 percent. If it contains a large proportion of lime, it becomes yellowish on firing. There is a good proportion of lime in all the samples. There is the possibility that the yellowish colour is masked by the red due to the presence of $Fe_2O_3$.

With regard to the proportion of alumina, iron, lime and magnesia, M13 exhibits abnormal values as compared with the corresponding values for the other samples.

The normative composition of the clays for the different samples of pottery has been worked out, and the values are given below.

From the normative composition, it will be seen that quartz varies in value from 30.30 percent to 36.18 percent. The values for the Periods I and II are somewhat uniform in their respective ranges. Albite, anorthite, magnetite, corundum and hypersthene vary in percentage from 16.40 to 24.46, 3.02 to 4.87, 5.81 to 10.20, 11.47 to 13.40 respectively.

The values for quartz are almost uniform in all the samples. The abnormal values for albite occur in M1, M2 and M13. The abnormal values for anorthite occur in M1, M2, M13, M16 to M18. The samples have more or less uniform values for magnetite, and hypersthene. The ab-

| *Samples* | *Quartz* | *Orthoclase Albite* | *Anorthite* | *Magnetite* | *Corundum* | *Hypersthene* |
|---|---|---|---|---|---|---|
| *M1* | 36.18% | 9.45% | 16.40% | 3.94% | 10.20% | 12.06% |
| *M2* | 33.90 | 9.45 | 19.74 | 3.71 | 8.47 | 13.31 |
| *M3* | 33.90 | 11.12 | 21.13 | 3.25 | 7.34 | 13.23 |
| *M4* | 34.26 | 11.12 | 19.74 | 3.48 | 7.96 | 13.40 |
| *M5* | 33.36 | 11.12 | 21.96 | 3.02 | 6.83 | 13.36 |
| *M6* | 31.74 | 11.68 | 23.07 | 3.71 | 7.35 | 12.64 |
| *M7* | 31.38 | 11.68 | 23.07 | 3.48 | 6.73 | 13.07 |
| *M8* | 33.06 | 11.12 | 20.57 | 3.48 | 8.47 | 12.64 |
| *M9* | 31.92 | 11.12 | 22.80 | 3.94 | 7.04 | 12.87 |
| *M10* | 31.20 | 10.56 | 22.80 | 3.71 | 6.53 | 12.94 |
| *M11* | 30.30 | 11.12 | 23.07 | 4.41 | 6.53 | 12.57 |
| *M12* | 30.78 | 11.12 | 23.35 | 4.41 | 6.12 | 11.94 |
| *M13* | 35.34 | 8.34 | 18.07 | 3.71 | 9.49 | 12.65 |
| *M14* | 31.03 | 11.12 | 12.80 | 4.87 | 6.53 | 11.64 |
| *M15* | 30.78 | 11.12 | 24.46 | 4.18 | 5.81 | 11.47 |
| *M16* | 35.16 | 10.56 | 18.63 | 4.64 | 8.87 | 11.75 |
| *M17* | 35.76 | 10.00 | 17.51 | 3.51 | 9.28 | 11.95 |
| *M18* | 34.74 | 10.56 | 17.51 | 4.87 | 9.49 | 12.45 |

normal values for corundum occur in M1, M2, M15 etc. In Period II, M13 exhibits abnormal values for quartz, albite, anorthite, magnetite, corundum etc.

## TEMPERATURE OF FIRING OF THE POTTERY

In the kiln, the clay first loses free and combined water. This process is very slow. At 800° C., it shrinks, and above 1000° C., reaction occurs with the formation of a mixture of silica and christobalite and multite.

Each sample of potsherd was broken into a number of fragments. They were kept in an electric furnace where the temperature could be accurately controlled. One fragment was heated to 400° C., another to 500° C., a third one to 600° C. and so on up to 1250° C., and their respective colours noted. As the fragments were heated, they slowly underwent a change in colour. The original colour disappeared at higher temperatures. These series of experiments reveal the maximum temperature up to which the potsherd may be heated without the original colour changing. This is also the temperature up to which it was originally heated by the potter in all probability.

Thus, in the case of these potsherds, the approximate temperature of firing is as follows:

| | |
|---|---|
| *M1* | 500–600 deg C |
| *M2* | 500–600 |
| *M3* | 500–600 |
| *M4* | 600–700 |
| *M5* | 600–700 |
| *M6* | 600–700 |
| *M7* | 600–700 |
| *M8* | 700–800 |
| *M9* | 700–800 |
| *M10* | 700–800 |
| *M11* | 600–700 |
| *M12* | 600–700 |

| | |
|---|---|
| *M13* | 600–700 |
| *M14* | 600–700 |
| *M15* | 600–700 |
| *M16* | 500–600 |
| *M17* | 500–600 |
| *M18* | 500–600 |

Since the ranges of temperature are not high enough, glazing of the pottery is to be ruled out.

From a study of the approximate temperature of baking, three groups of wares can be traced in Period I. M1 to M3 have been baked up to 500–600° C. M4 to M7 have been baked up to 600–700° C. M8 to M10 have been baked up to 700–800° C.

The wares of Period II have a uniform temperature range of baking, namely, 600–700° C. The wares of Period III have also got a uniform temperature-range of baking. Thus, taking all the samples, most of them have been baked in the region of 500–600° C. or 600–700° C. A few of them have been baked at a maximum temperature range of 700–800° C.

These studies also indicate the "effective" temperature which could have been reached in a kiln of the type used in ancient times in the region of Maski. They had no other fuel probably except wood. They could not have had other sources of power, whereby a higher effective temperature could have been reached. Some of the samples have been baked in a higher temperature range. This has probably been effected through the wares' being placed more in the central portion of the kiln.

On breaking any of the potsherds, the broken section reveals a dark band of partially baked or almost unbaked clay in the middle. Whenever such a dark band occurs, it is a clear indication that the temperature range of baking could not have been higher than the ones specified. At higher temperature range, the section is completely baked and there is no dark band. This fact has been verified experimentally with many of the potsherds under investigation.

The fragments of the potsherds were also heated in a heating microscope and a series of photographs taken while the materials were being heated to various temperatures. In the series of photographs, the temperature at which the material begins to soften can be easily noted. This occurs at about 1000° C. for all the samples.

In computing the temperature of baking, one of the objectives is to account for the colours of the potsherds. The following variations occur among the potsherds made available for study: grey, pinkish buff, black-and-red, russet-coated.

The grey is accounted for by the presence of organic matter. Where the colour persists in spite of its being heated in an oxidising atmosphere, it must be due to the presence of manganese.

Pink and red are mere variations and are accounted for by the presence of iron in the clay which imparts the reddish tinge. The black-and-red coloring is due to the presence of iron in the clay. If the baking is done in a reducing atmosphere, the pottery becomes black. An oxidising atmosphere gives red ware. The reducing atmosphere can be created by (a) putting organic matter in the kiln, (b) covering portions of the pottery (purposely or not) from the effects of the oxidising flame, (c) arranging the pieces of pottery one over the other in an inverted manner—technically called "inverted firing"—so that portions of one piece of pottery, which is covered by another, is black after firing on account of the effective prevention of the oxidizing influence. That a black piece has been baked under reducing conditions may be proved by heating the black potsherd in an oxidising flame. It turns red. Conversely, when a red potsherd is heated in a reducing atmosphere, it turns black. If a large amount of black-and-red ware is available from the same excavation, the question of "inverted firing" may be resolved experimentally.

It has been reported that there are samples of painted ware from Maski excavations. It is also stated that the painted ware is an under-slip decoration with lime, which has been fired subsequently without, however, resorting to "biscuit firing" as is usual with decorative wares in porcelain. Since others have been engaged in this study, it has not been taken up for investigation here.

I must express my thanks to Mr. A. V. Krishnamurti who helped me in the experimental investigations.

# Methodological Problems of $C^{14}$ Dating

By Elizabeth K. Ralph
University of Pennsylvania

ABSTRACT

From the brief summary of Carbon-14 dating techniques presented, it is apparent that the proportional gas-counting methods continue to be the most practical for routine dating. The problem of obtaining a reliable $C^{14}$ date is twofold. The first is primarily that of dating a representative sample —one which is, without doubt, contemporary with the culture or structure to be dated. The second is due to small uncertainties in the fundamental premises of the method— due to variations in atmospheric $C^{14}$ inventory in past times and the uncertainty in the determination of an "effective" half-life value for dating. Measurements of samples of known age, as determined by dendrochronology, provide a means of elucidating the latter problem.

It is appropriate to talk on this subject to members of the Division of History of Chemistry of the American Chemical Society because this dating method was envisaged and developed by a chemist—namely, Prof. Willard F. Libby; and naturally, the techniques employed are mainly chemical. The explanation of the existence of natural C-14 is now an old story—to wit, its production as a result of collisions in the upper atmosphere between $N^{14}$ atoms and neutrons from cosmic rays. The resultant $C^{14}$, in an equilibrium state between production and decay, constitutes approximately one part in $10^{12}$ parts of our atmospheric $CO_2$.

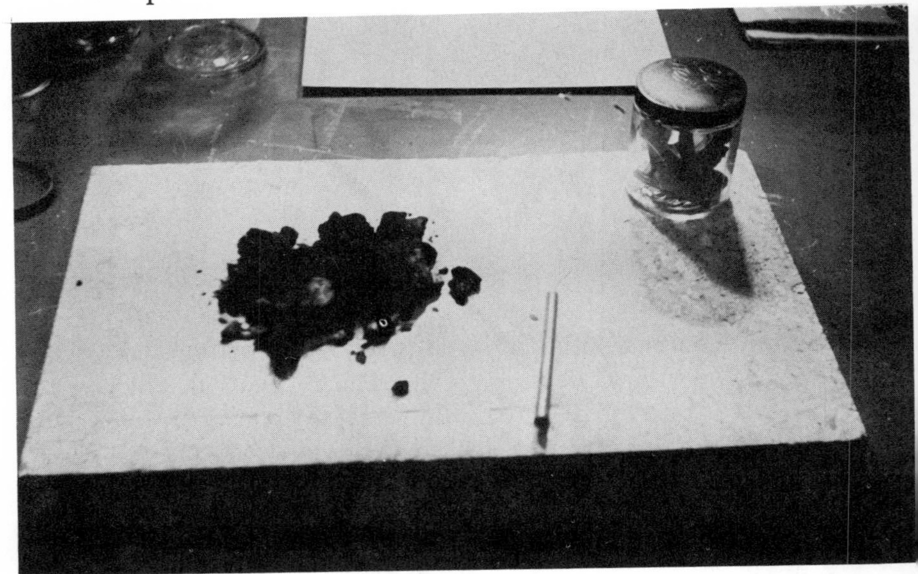

**Fig. 1. Sample—charcoal.**

For the counting of this extremely low level radioactivity, the solid carbon geiger method which Prof. Libby perfected has now been replaced in most laboratories by gas-counting techniques. This was necessitated in 1954 and 1955 by the susceptibility of solid carbon to contamination from fall-out and other radioactive by-products. Another drawback was that due to self-absorption in counting solid carbon, only 5 percent of the radiations were detected. Gas counting has several advantages—notably, approximately 100 percent counting efficiency, and with pure $CO_2$ (the method that most laboratories have adopted) the fact that virtually no isotopic fractionation is introduced in the processing. The impurity requirements are stringent in that electronegative ions must be reduced to less than one part per million for $CO_2$ to be a good counting gas in a proportional counter. Better counting gases such as acetylene and methane are used by some laboratories. The latter, especially, is increasing in popularity because of recent improvements and simplifications in the technique of conversion of $CO_2$ to $CH_4$. Here again, $CO_2$ has an advantage in that it is a poorer detector of neutrons, and background variations due to fluctuations in the cosmic-ray intensity as a result of daily changes in the barometric pressure are minimized.

It is suggested frequently, mostly by eager salesmen, that scintillation counting would simplify our problems. With the techniques available at the present time, this is not the case. Since natural $C^{14}$ is weak in two respects—the beta particles are emitted at the slow rate of 14 dpm/g for modern carbon and their average energy is 50 Kev, a liquid scintillator must be used which contains a large proportion of the sample to be counted. The choices available are to convert the sample to a suitable solvent such as benzene or toluene or to make simpler compounds such as alcohol and use these as dilutents with consequent reduction in counting efficiency. (In the latter case, no gain in counting rates over gas counting is achieved.) Both have been tried successfully, but as far as I know, neither has been continued long for routine dating. The real drawback is that the more elaborate chemical procedures required for conversion to organic compounds are just too much more trouble when one is faced with the processing of several hundred samples per year. Additionally, there is the more serious disadvantage that the reactions involved do not go to completion and fractionation may be introduced.

**Fig. 2. Chemical train, front view.**

In the first series of figures the apparatus which we use at the University of Pennsylvania for the conversion and counting of $CO_2$ is shown.

To use this isotope for dating, the first assumption required is that the mixing rate of the atmosphere be rapid. The world-wide tracing of fall-out from bomb tests and measurements of the increase in the atmospheric $C^{14}$ in the

Southern Hemisphere following the first large-scale testing in the Northern have provided data that indicate that the rate is of the order of 2 years, a rate which is sufficiently

**Fig. 3. Chemical train, CaO.**

rapid in view of the average life of a $C^{14}$ atom of approximately 8,000 years.

**Fig. 4. Shields and counting equipment.**

The first requirement of a sample for dating, therefore, is that it contain a representative amount of $C^{14}$, that is, that the sample to be dated grew, then died (and ceased to incorporate $C^{14}$) during the period that it is believed to represent. From this consideration it is apparent that only the outer growth layers of a tree will represent the time of the cutting of the tree and usually the date of construction of a building. The inner layers, as soon as each one becomes inner, have ceased to take in $C^{14}$. For this reason, too, samples of grain, cloth, hide, and similar materials are ideal because they were used, and presumably buried in anticipation of the present-day excavator, at the time they ceased to incorporate $C^{14}$.

Another problem, also one of the true contemporaneity of the sample, is that of so-called humic contamination. This may occur from the adsorption of younger finely divided, slightly soluble organic matter by the sample, and is most to be suspected when a sample has been exposed to continual washing by ground waters. It has not been found to be a serious contaminant except in very old age-ranges, that is, tens of thousands of years. We believe that most of this humic acid is removed by pretreatment in a

**Fig. 5. Sequoia section in lab.**

hot solution of NaOH following the usual HC1 treatment for removal of inorganic carbon compounds.

**Fig. 6. Bristlecone tree.**

Now that we know that the mixing rate is rapid and that we have selected representative samples, we come to the most serious methodological problem. Since the production of natural $C^{14}$ is dependent upon the number of neutrons present in the upper atmosphere, it is necessary to assume that the cosmic-ray intensity has been constant during past eras, that the earth's magnetic field has not changed appreciably, and that the equilibrium between atmosphere, oceans (the largest reservoirs of carbon), and the smaller bodies of water has remained steady. Libby demonstrated in 1949 [1] and again in a recent paper [2] that this assumption is correct within 10 percent, but with the increased precision of counting techniques, there are indications [3] that the $C^{14}$ content of the atmosphere may have differed in some past time intervals by 3 percent or more and/or that the previously accepted half-life value (5568 ± 30 years)[4] is not the true one.

Indeed, at the recent radiocarbon conference held in Cambridge, England in July, 1962, three new half-life determinations were reported. They are as follows: At Uppsala, Sweden by Karlen, I and Olsson, J., a value of 5680 ± 40 years; at the NBS by Mann, W. B., 5740 ± 40; In the United Kingdom at Aldermasterton, by H. Wilson, 5780 ± 50.

**Fig. 7. Bristlecone tree, section cutting.**

The average of these three is 5730 (± 40) and this value was adopted by the conference as the half-life of $C^{14}$. However, it was felt that more attempts to measure this half-life should be made, especially, by different techniques. The determinations were all similar—made by proportional counting of $CO_2$ (except for the NBS one which was counted also in geigers) and all were subject to the same types of errors—mass spectrographic measurement, dilution, subsequent absolute counting, and the error which is most difficult to assess—namely, adsorption of some of the $C^{14}O_2$ along the way. For this last reason, the consensus at the radiocarbon conference was that these half-life values may, if wrong, be too low. In spite of this, the conferees in Cambridge voted to continue to calculate radiocarbon dates with the Libby value, 5568 years. The change to the new half-life, if adopted, would cause an increase in $C^{14}$ ages of 3 percent.

The problem, for dating however, is a complex one and its nature is revealed by the measurements of samples of known age. First of all, these, themselves, are somewhat of a problem. Many of the archaeologically-dated samples are subject to the outer-growth error; that is, they consist of large beams which have been trimmed for construction; and even such familiar chronologies as that of the Egyptian dynasties are subject to uncertainties beyond approximately 1900 B.C. We and several other laboratories have, therefore, turned our attention to sections of wood dated by dendrochronology.[5] The sequoias (*Sequoia gigantea*) provide samples extending back 3000 years and it is hoped that another 1000 years or more will be attained by the Bristlecone Pines (*Pinus aristata*) when the tree-ring dating of these has been completed.

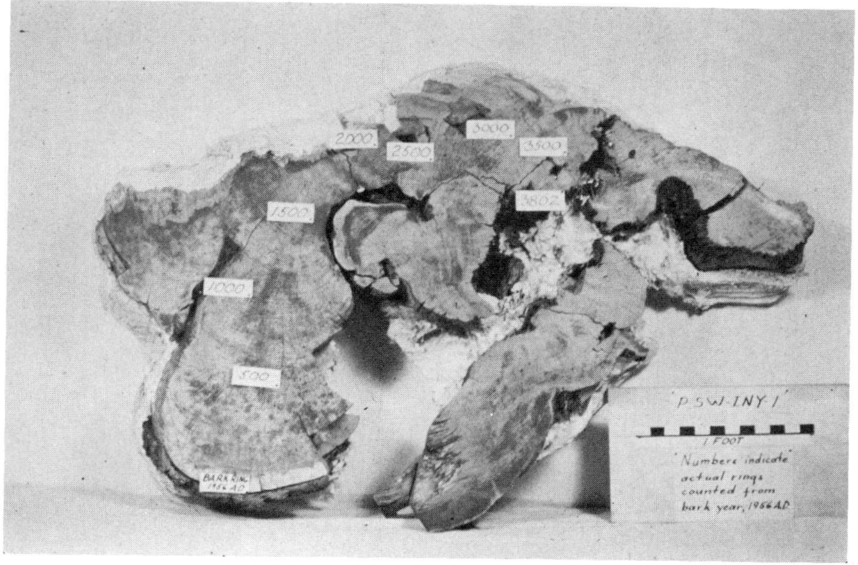


**Fig. 8. Bristlecone section in lab.**

A few examples of both types of trees, and the results of the $C^{14}$ measurements are shown in Figs. 5–9.
It can be seen from Fig. 9 that there was a definite increase in the atmospheric $C^{14}$ inventory in the period of A.D. 1700 to 1500. Then, we have a scatter in our data in the period A.D. to 500 B.C., followed by a general upward trend from 600 B.C. back in time. Here, it is apparent that the new half-life value, or even our effective value of 5800, which is the average obtained from our measurements of tree-ring dated samples, are in better agreement with known ages. As is evident, however, from Fig. 9, the problem is complex and for accurate dating in historical time periods, effective half-life values will be necessary; and beyond this,

our best hope will rest on an average value based on more data than we now have.

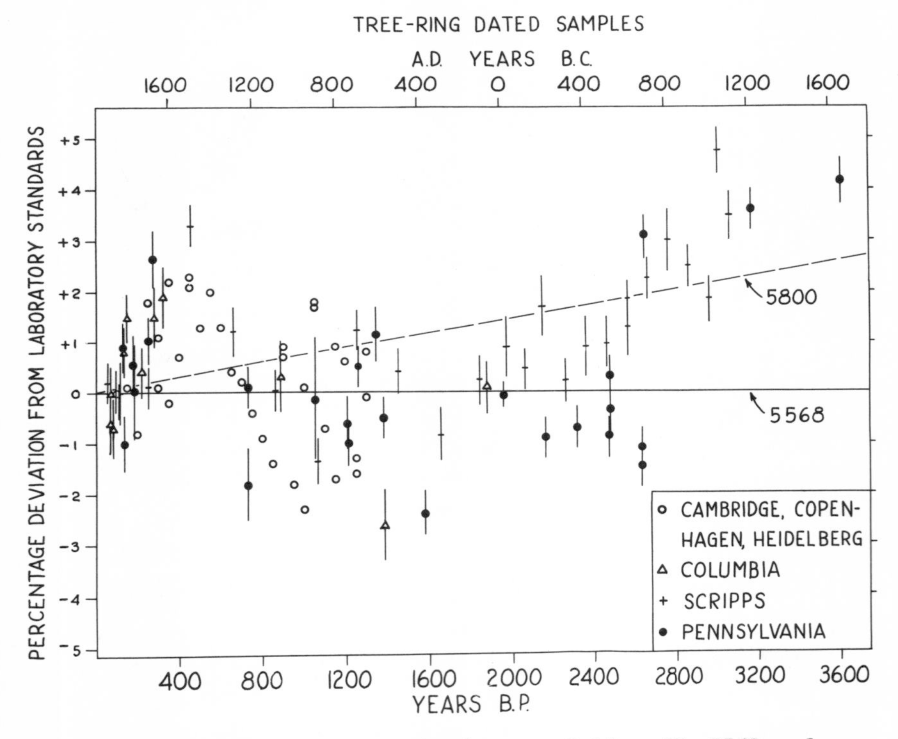


**Fig. 9. Tree-ring dated samples by several labs with 5568 and 5800 half-lives shown.**

These problems are not serious for geological dating for which a consistent time scale in radiocarbon years is adequate. It is when our dates overlap historical periods that additional accuracy is needed. Two examples of these stringent requirements are discussed in the reports of our dates for Tikal, Guatamala (the Maya calendar correlation problem)[6] and for sites in the Mėditerranean area.[7]

In conclusion, we find that even though we can now count natural $C^{14}$ with an accuracy of 0.5 percent or less, we have not yet solved our methodological problems, *but* we are aware of the discrepancies. We hope to reduce them and to clarify their causes in the near future.

## NOTES

1. Libby, W. F., Radiocarbon Dating, (2nd ed., Chicago 1955) p. 10.
2. Libby, W. F., The Veracity of Radiocarbon Dates (Paper for the volume in honor of Harold C. Urey on his 70th birthday, 1962).
3. Ralph, E. K. and Stuckenrath, R., "Carbon-14 Measurements of Known Age Samples," *Nature* 188 (Oct. 15, 1960) 185–87.

Willis, E. H., Tauber, H., and Munnich, K. O., "Variations in the Atmospheric Radiocarbon Concentration over the Past 1300 Years," *American Journal of Science Radiocarbon, Supplement 2* (1960) 1–4.

Broecker, W. S., Olson, E. A., and Bird, J., "Radiocarbon Measurements on Samples of Known Age," *Nature* 183 (June 6, 1959) 1582–84.

Vries, Hl. de, "Variations in Concentration of Radiocarbon with Time and Location on Earth," *Proc. Kon. Ned. Akad. van Wetenschappen*, B 61 (1958) 1–9.

4. Libby, W. F., *Radiocarbon Dating,* (2nd ed., Chicago, 1955) p. 36.
5. This work is supported by Grant No. G-14094 from the National Science Foundation.
6. Satterthwaite, Linton and Ralph, E. E., "New Radiocarbon Dates and the Maya Correlation Problem," *American Antiquity* 26 (October, 1960).
7. Kohler, E. L. and Ralph, E. K., "C-14 Dates for Sites in the Mediterranean Area," *American Journal of Archaeology,* 65 (1961) 357–67.

# Some Notes on Pre-Columbian Metal-Casting

By W. C. Root
Bowdoin College

It is now evident that the methods used in the casting of objects of gold and other metals were essentially the same over the entire area occupied by the higher cultures of the Americas. Objects from different regions differ very greatly in style, but the methods used in their manufacture were almost identical. For this reason, data on casting obtained from the examination of a Mexican, a Colombian, or a Peruvian metal object can be assumed to be of general rather than local application.

The only early account of metal-casting that has come down to us is that written in the sixteenth century by Friar Bernardino de Sahagún from information given him by his Aztec helpers. This work has recently been translated from the Aztec by Anderson and Dibble.[1] Details of the casting and "gilding" of gold and tumbaga (an alloy of gold and copper) objects are given in Book IX, Chapter 16.

Dudley T. Easby, Jr., as a result of examining numerous cast objects of gold, has been able to clarify the account of

Sahagún, and in several recent articles has explained the methods used in admirable detail (2, 3, 4).

In this paper I will describe the results of my examination of five cast objects which amplify to some extent the conclusions reached by Easby.

The principal method of casting used in prehistoric America was that known as the "lost wax" process or *à cire perdue*. The first step was to make a core from a mixture of clay and charcoal which was close to the shape of the desired object. The core was then covered with a layer of wax which was modeled or carved to the desired shape. Details were frequently added by pressing on wax pellets or "wires." Sometimes these were so elaborate that the final product resembled filagree.

Small wax rods to provide air vents to permit the escape of trapped air and a larger wax rod to provide a pour for the molten metal were then attached to the wax model in appropriate positions. If the object was to be a closed casting like a figurine or statuette, thorns or small rods of wood were inserted through the wax into the core to support it when the wax had been removed.

The wax model was then treated with a slurry of powdered charcoal and water. This filled the tiniest interstices and incised lines, and resulted in the outer metal surface being an exact duplicate of the wax surface and so smooth that it needed little polishing. After drying a clay covering was placed over the model with a funnel-shaped opening to the wax pour.

After a thorough drying, the mass was gently warmed to melt the wax which was then poured out. The clay model was next placed in a furnace and heated to a high temperature and the molten metal tipped into the pour. After being slowly cooled, the clay covering and the core were scraped away and the core supports were pulled out. The metal in the air vents and the pour was cut off as close to the surface as possible. The holes left by the core supports and other unwanted openings were closed with thin metal plugs. Since the gold used sometimes contained copper, if the surface was dark the object was boiled in a solution of plant acids to remove the film of copper oxide,[2] and was then burnished.

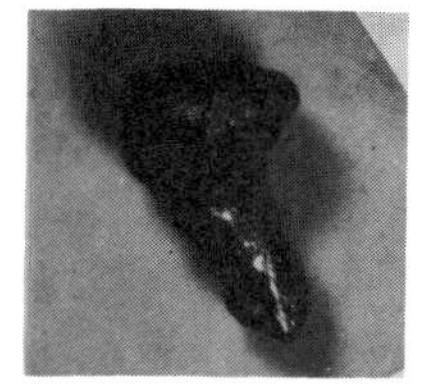

**a. (natural size).**

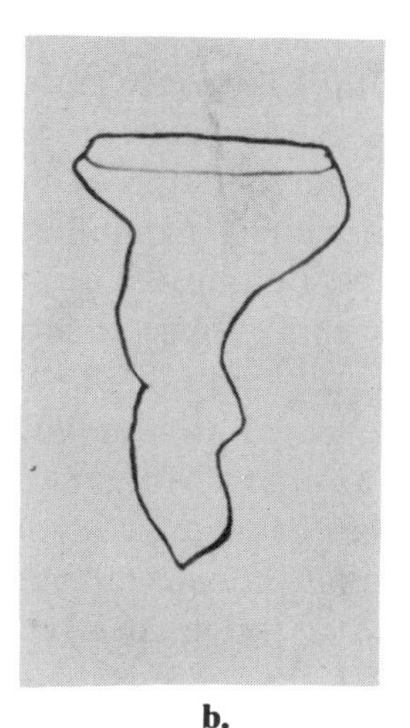

**b.**

**Fig. 1.**

In Figure 1 is shown a gold pour which had been removed from a casting.[3] It is 1.8 cm in length, the top is an oval of 1.4 x 1.1 cm, and the stem is about 1 cm long and 0.5–0.6 cm in diameter. The stem had been cut from opposite sides with a chisel and then broken off. If the top surface of the pour is assumed to be horizontal since it represents the surface of the molten metal, the stem slopes at an angle of roughly 60°. This confirms Easby's theory that the model was tilted to allow the metal to run down the cavities in the lower side while the air would pass through the openings in the upper side until it finally escaped by the air vent.

Three objects from Colombia are shown in Figures 2, 3, and 4. I had a chance in the winter of 1959–60 to examine the extensive collection of Colombian gold in the British Museum. Like most museums where the collections consist of gifts or purchases from dealers, most of the objects have little documentation as to provenance. These three objects are said to be from the Cauca valley and belong to the Quimbaya culture. They date from 900–1100 A.D.

The small figurine shown in Figure 2 is of a dull gold and is only 6 cm. high.[4] It appears to be a primitive version of the seated woman found in two splendid examples in the University Museum in Philadelphia.[5] Since the entire back has been cut away it is possible to see the entire inner surface of the ventral side of the figure, Figure 2 (c). The surface is rough and reproduces the outer surface of the core. The outer surface of the figure is quite smooth as it had received the wash of powdered charcoal. The core appears to have been carved very extensively so as to give a close approximation to the final form, notice for example, the mouth and the eyelids. Because of this the metal when cast is almost of uniform thickness. There are four holes for core supports—two in the knees and two on opposite sides of the neck where a rod had been pushed through from one side of the wax covered core to the other. Only one of these four holes still retains its plug. The outer surface is so well polished that it is impossible to tell where the pour or air vents were attached.

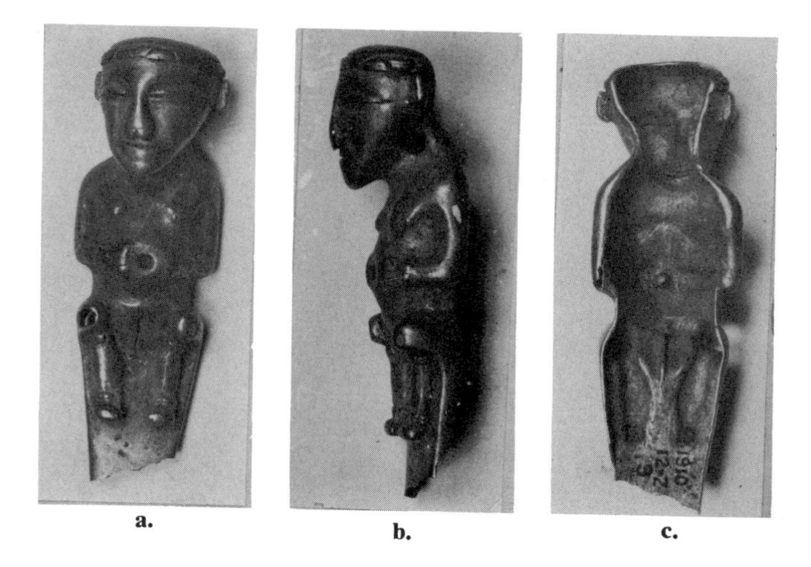


**Fig. 2. (natural size).**

The large male figure shown in Figure 3 weighs 320 grams and is 30 cm high.[6] It is made from a gold-rich tumbaga, "red gold," and contains 58 percent gold. It shows yellow spots so was probably lightly gilded by the process of *mise en couleur*. The dorsal and ventral surfaces have been forced together as if the figure had been crushed by a heavy weight. There is a large hole in the upper right arm and numerous large cracks. It is odd that the forearms holding the spirals (flowers?) were not also crushed.

Like the hollow figure described by Easby, the lower legs and feet were cast separately and were soldered on to the upper legs just above the knee bands. It is not possible to tell if the hands and spirals were also separately cast and

a.

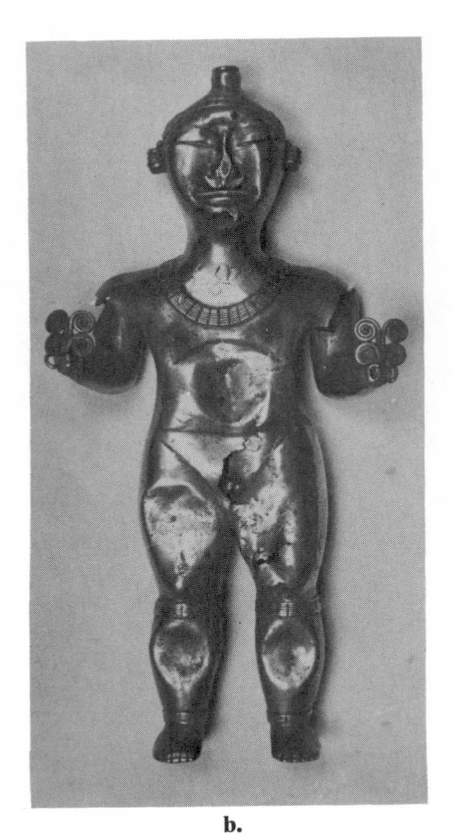

b.

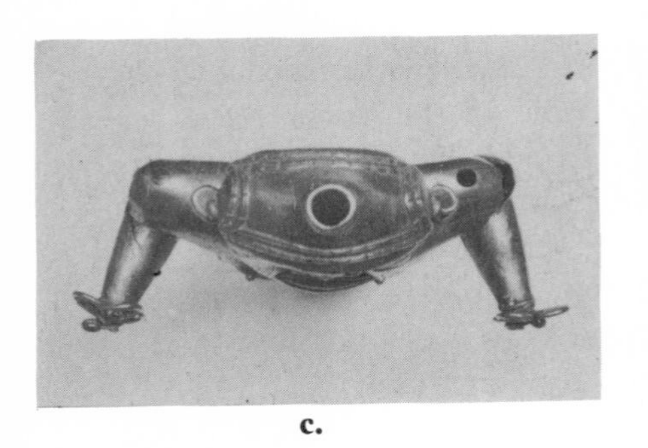

c.

d.

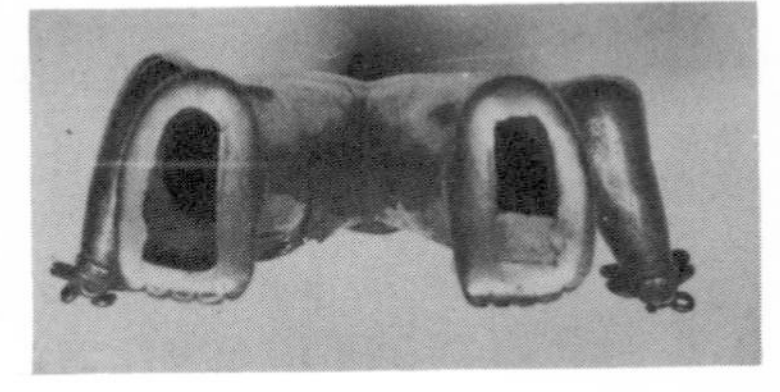

e.

Fig. 3.

soldered on to the arms at the wrist, although this is probable. It is 20 cm from the top of the head to the knees. This distance probably represents the length of the largest casting that could be conveniently made in the furnaces that were ordinarily available.

Core supports had been inserted in the two shoulders and in the navel. The only one that retains its plug is the one in the right shoulder. Because of the large hole in the right arm it was possible to observe the underside of this plug. The plug is shown in outline in Figure 3 (d). The opening, 5 mm in diameter, has a rim around it on the inner side. The core must have been cut away in order that the rim could be formed by the wax. The rim would be of help in keeping the plug in place. The plug in this case was somewhat thinner than the metal of the shoulder. Attached to the inner surface of the plug was a hemisphere of dark material, probably similar in composition to the charcoal-clay mixture used for the core. The rod to be used for the core support may have been pushed into a small ball of the core material, then a little molten metal poured into the depression. After cooling and the cutting away of excess clay one would have a plug that would exactly fit the hole to be plugged with enough adhering clay to keep it firmly in place inside the little rim.

The head-dress, necklace, wrist, knee and ankle bands were made by applying a thin wax ribbon about 1 cm wide to the wax coating. Then the ribbons were decorated by scoring them with a sharp instrument. The ear ornaments and spirals held in the hands were made by the application of wax "wires."

The third object from Colombia is the curious ornament shown in Figure 4[7]. It is of gold, 6.2 cm long, with 14 elements made up of tubes with an outside diameter of 4.4

mm. Probably the artisan first rolled out some of the core material into a rod about 3 mm in diameter. He then coated this with a layer of wax about 1 mm thick. Then the wax-coated rod was cut into pieces 5 mm in length and these were pushed together side by side until they stuck together. Then two incised lines were made the length of the ornament on each side. The casting was then completed in the usual manner.

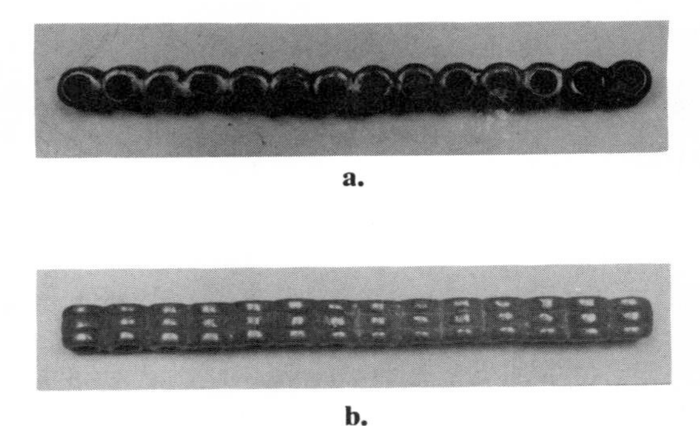

a.

b.

**Fig. 4. (natural size)**

The last object to be discussed is shown in Figure 5.[8] It is a fragment of a "turtle" necklace, one element of which is partially shown in Figure 5 (a). Three loops and the attached ring are shown in Figure 5 (b). The loops are 1.2 mm in diameter and the ring is 6 mm in diameter. This necklace is a typical example of Mexican metalwork from Oaxaca where "wirework" and "false filigree" were frequently used for ornamentation. The fragment is dark grey in color with a metallic lustre. It contains 94 percent copper and 6 percent tin and lead. It probably dates from 1300–1400 A.D.

a. (natural size).

a.

b.

c.

d. Etched with $H_2O_2$ and $NH_4OH$.

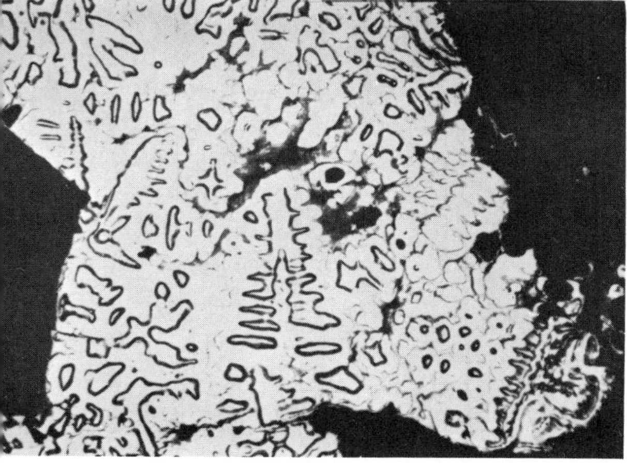

Fig. 5.

It is a common belief that in ornaments such as this the delicate decorations could not possibly have been cast but must consist of fine wires soldered to a metal base. The only way in which this hypothesis can be tested is to section such an object and subject the section to a metallographic examination. Most museum directors object to having their objects cut into pieces even though they may be hidden away in closets where no one sees them, so the only chance one has for a metallographic examination of an object is when it is so badly damaged it cannot be exhibited to the public. Fortunately, such a fragment was available. It is to be hoped that eventually a sound gold object can be examined.

The fragment of the turtle necklace was sent to Dr. Morris Cohen of the Department of Metallurgy of the Massachusetts Institute of Technology. He sectioned it through the ridge at the top of the loops, along one of the loops, and through the attached ring at the bottom. Figure 5 (c) shows a photomicrograph of the section along A-B, and Figure 5 (d) shows a photomicrograph (enlarged 100 times) of the junction of one arm of the loop with the suspension ring. Dr. Cohen reported that the whole section was cast in one piece with no sign of soldering or welding at the joint. The cast structures continue through the joint without a break. This is also true of the junction where the top of the loop joins the band at A. In the photomicrographs, the very black portions are due to corrosion products of the copper.

These results, and those of the few other metallographic studies that have been made of objects with wirework decorations, show that, contrary to popular belief, most if not all of the "false filigree" and "wirework" decorations on pre-Columbian metal objects were cast at the same time as the body of the object and were not soldered on after casting. More metallographic examinations of such objects are needed to prove this conclusively, but this is not likely until someone is allowed to mutilate a really first-rate gold ornament.

There are still aspects of casting that are obscure, such as the method by which the molten metal was added and the nature of the furnaces that were used, but the important details of the methods that were used by the pre-Columbian metal workers seem to be those described by Easby. Other metal objects will be examined but it is unlikely that other than minor variations in the procedure described by Sahagún will be found.

## ACKNOWLEDGMENTS

I wish to thank Mr. Adrian Digby, Keeper of the Department of Ethnology of the British Museum, and Mr. C. A. Burland, for their help in making it possible for me to examine and photograph the pre-Columbian metal objects in the collections of the museum. I also wish to thank Professor G. Bing, Director of the Warburg Institute, for providing me with office space and the facilities of their library, Dr. Gordon Ekholm of the American Museum of Natural History for permission to section a portion of the "turtle" necklace, and Dr. Samuel Lothrop of the Peabody Museum at Harvard for calling to my attention his unique example of a "pour."

## NOTES

1. Dibble, Charles E. and Anderson, Arthur J. O. "Florentine Codex," *Monographs of the School of American Research,* Number 14, Part 10, 1959, pp. 73–78.

2. Easby, Dudley T. Jr., "Sahagún Reviviscit in the Gold Collections of the University Museum," *The University Museum Bulletin,* University of Pennsylvania, Philadelphia. Vol. 20, 1956, pp. 3–15.

3. Easby, Dudley T. Jr., "Ancient American Goldsmiths," *Natural History,* New York. Vol. 65, 1956, pp. 401–409.

4. Easby, Dudley T., Jr., "Sahagún y los Orfebres Precolombinos de Mexico." *Anales del Instituto Nacional de Antropológia e Historia,* Mexico City. Tomo IX, 1957, pp. 85–117.

5. Boiling the object in an acid solution was also the first step in one method of "gilding" a tumbaga object by the process of *mise en couleur*. Repeated heatings, boilings to remove the oxide, and burnishings eventually resulted in a surface of almost pure gold.

6. From the collection of Samuel K. Lothrop.

7. British Museum, catalogue number 1910 12–2 9. "From Quindio."

8. Easby, *op. cit.*

9. British Museum, catalogue number 89 10–1 1. "Colombia—like ones from a grave at Salento, Cauca, the country of the ancient Carrapas."

10. British Museum, catalogue number 1910 12–2 15. "From Quindio."

11. American Museum of Natural History. Loan. "Mexico."

# Some Materials of Glass Manufacturing in Antiquity*

By Edward V. Sayre and Ray W. Smith
Brookhaven National Laboratory, Upton, N. Y.
and
International Committee on Ancient Glass, Dublin, N. H.

The chemical analysis of ancient artifacts can reveal much about the methods and materials used in their manufacture. Analysis of ancient glass in particular has shown that in different periods and regions sometimes quite different ingredients and formulations were used in its production. For example numerous analyses have established that prior to about the tenth century A.D. the glass produced in the areas of North Africa, Western Asia and Europe was predominately a soda-lime glass to which lead oxide was only occasionally added. At approximately the end of this period, some outright lead glass was produced in Islamic areas. Following the tenth century in Northern Europe potassium rather than sodium became the predominant alkali in glass. This change was first explored extensively

* Research performed in part under the auspices of the U. S. Atomic Energy Commission.

by Geilmann [1] and his co-workers who also observed accompanying changes in manganese and phosphorus concentrations and concluded that the introduction of beechwood ashes as an alkali was responsible for the change. Bezborodov [2] and others have demonstrated a predominately potassium alkali was used in glass produced in Russia during the eleventh to thirteenth centuries and that additional high-lead glasses were produced there.

Within these basic formulations, subordinate variations in composition are often encountered. These differences in part might reflect characteristic compositional differences in particular raw materials when obtained from different geographic regions. Such regional variation can be very helpful in indicating the origins of these components. Occasionally chronology can be established if the composition of raw materials for various reasons changed with time. Additional significant variations within basic compositions can be traced to the deliberate addition of materials to affect properties such as color, transparency, fluidity at higher temperatures, etc.

At Brookhaven National Laboratory, we have now completed the analysis of more than four hundred glasses of the early, occidental "soda-lime" type. In the selection of specimens, we attempted to sample reasonably comprehensively glass production throughout Europe, North Africa and Western Asia as far east as Afganistan and India from the middle of the second millennium B.C. until roughly the twelfth century A.D. The Far Eastern lead-barium glasses [3] and the more recent European potassium rich and high-lead glasses have been excluded. Therefore, except for a few lead-rich Islamic glasses, these specimens have a basically similar composition. Within this approximate uniformity, however, some geographic and chronological variation and some differences that appear to reflect technological changes have emerged. The present discussion will be concerned with some of the factors that appear to relate to basic technical formulations and to materials knowledgeably added in given instances to produce desired results.

## DISTRIBUTION ACCORDING TO CONCENTRATION OF MAGNESIUM

It has been noted for some time that certain ancient glasses possess notably more magnesium than others. The magnesium-rich glasses have sometimes been called "dolomite" glasses upon the premise that dolomitic limestone had been used in their formulation. However, in our specimens we have observed a pronounced correlation between the concentration of magnesium and some of the alkali metals, but essentially no correlation between concentrations of magnesium and calcium. A high magnesium concentration is accompanied by a relatively high potassium concentration. Taken as a whole the high magnesium specimens have slightly less sodium so that the total alkali content in high- and low- magnesium glasses is somewhat the same. Such comparisons are shown in Fig. 1. For each element the distribution solidly blocked in is that of the same high-magnesium specimens, the light distribution that of the specimens with low concentrations of magnesium, and the crosshatched area is a region of overlap of the two distributions.

The magnesium concentrations divide into two almost completely separate ranges. The very few specimens of intermediate magnesium composition could be unambiguously classified as to type upon the basis of their potassium concentrations. In the case of potassium, the two distribu-

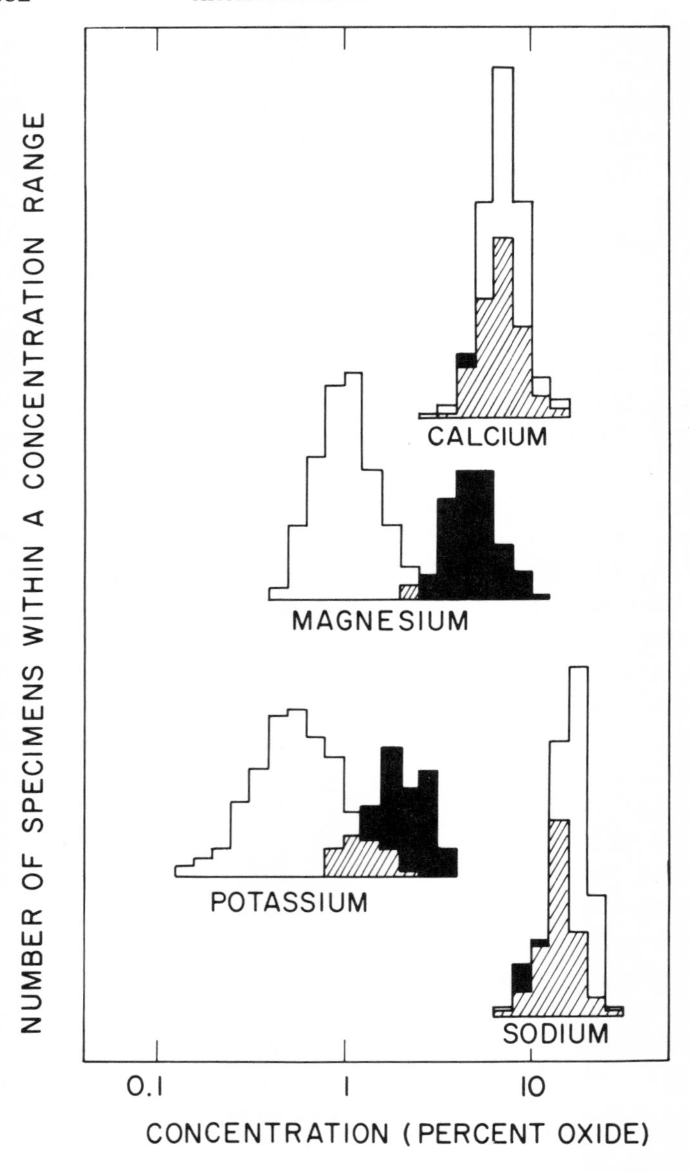


**Fig. 1. Comparison of the occurrence of calcium, magnesium, potassium and sodium oxides in the high-magnesium glasses, darkened distributions, and low-magnesium glasses, light distributions. The regions in which two distributions overlap are crosshatched.**

tions are slightly more broad and the region of overlap, the crosshatched area, greater. Fig. 1 also makes apparent the extent to which the calcium distributions for the two types of glass nearly superimpose and shows that the sodium distribution for high-magnesium glasses is slightly shifted to a lower range than that of low-magnesium glasses.

The most basic conclusion to be drawn from these data is that indeed two statistically distinct types of glass were produced throughout this period and region. Glass of the second millennium B.C. was of the high-magnesium type. Specimens datable from about 1500 to 800 B.C. excavated in Egypt, Mycenaean Greece, Crete, Anatolia, Mesopotamia and southwestern Persia have had typical high magnesium composition. Following this period and continuing until the latter part of the first millennium A.D. the low-magnesium formulation replaced the high in most of these areas. However, there is strong evidence to indicate that high magnesium glass continued to be manufactured in the Tigris-Euphrates region and in India. Smith [4] has described the occurrence of these two basic glass types in considerable detail. The manufacture of low magnesium glass also spread into Europe. In locations as far north and west as Gaul, Scandinavia and England it has been found with essentially the same basic composition as Middle Eastern glass of the Roman period. The low-magnesium formulation was retained in European glass manufacture until the use of wood ashes was introduced in about the tenth century. Also by about the tenth century many of those Islamic areas which had adapted the low magnesium composition were returning to the production of high magnesium glass. Some indication that Venice eventually followed Islam in this tradition is provided by an analysis we have made of a fragment of sixteenth to seventeenth cen-

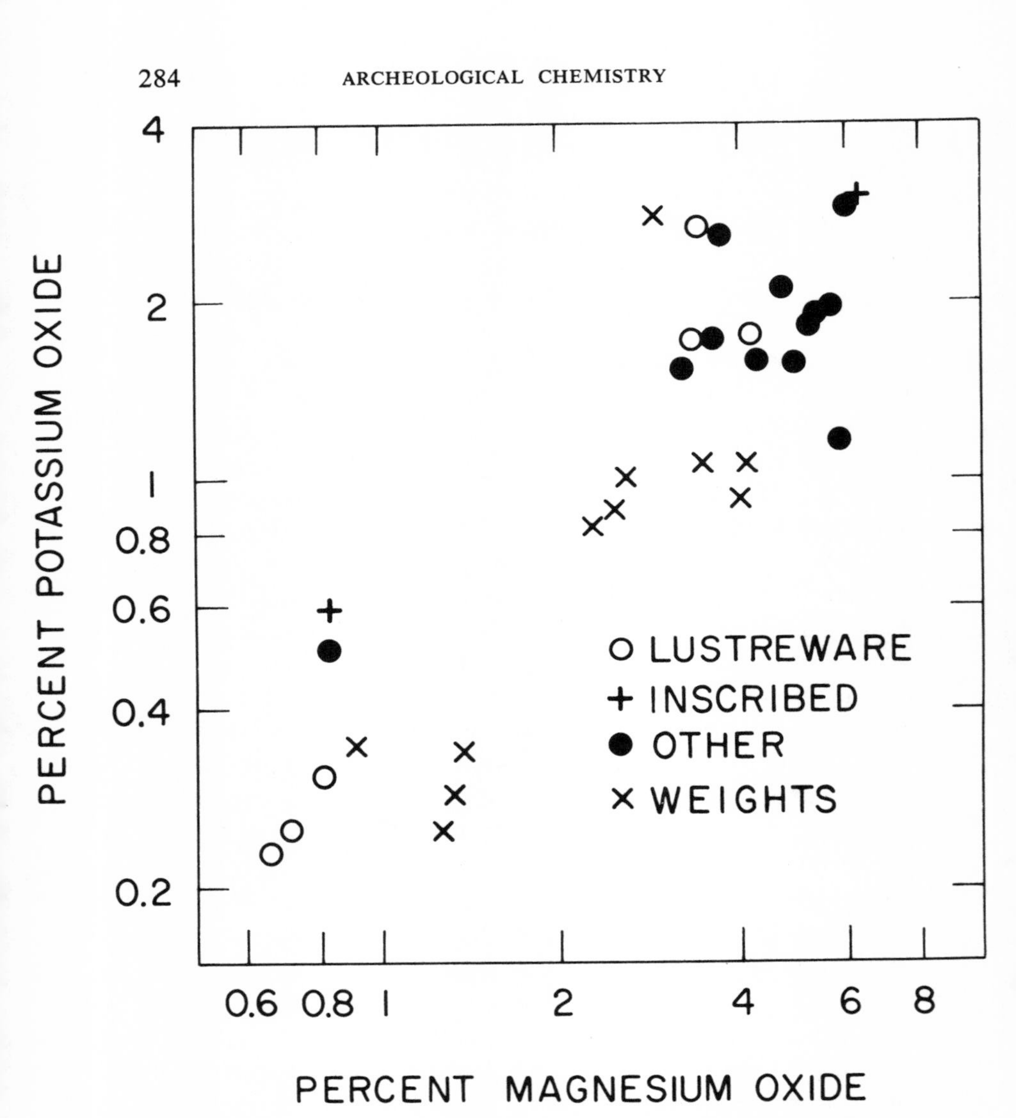


**Fig. 2. Occurrence of potassium and magnesium oxides in glass found in Egypt or neighboring areas from the early Islamic centuries.**

tury Venetian goblet. It proved to have typical high magnesium composition.

It was more usual for only one or the other of the two types of glass to be standard in a given area at a given period. There were, however, regions where, at least at certain times, both types were produced. Egypt is interesting in this respect in that its earliest glass was of the high-magnesium type, then from the middle of the first millennium B.C. until Islamic times the standard composition appears to have been low magnesium, and eventually during the early Islamic centuries its glass included both varieties. Fig. 2 shows the occurrence of both types of glass attributed to the Islamic centuries and found in Egypt. In this figure, magnesium and potassium concentrations of these specimens are plotted in correlation to each other. It not only shows that the specimens separate into two distinct groups with high and low concentrations of these elements, but emphasizes the relation that exists between the two elements. Two of the specimens bore cartouches in molded relief containing identical Cufic inscriptions that would appear to relate them closely. However, as can be seen in Fig. 2, one is clearly of the high magnesium formulation and one of the low. Glass weights also exhibit both compositions, those weights with high-magnesium composition being of a later date. Some other related groups which are split between the two glass types are three lustreware fragments with low magnesium and three with high. There are stylistic differences between these two groups, however, and there may be a small chronological separation in this instance.

The Tigris and Euphrates valleys form another region in which both types of glass apparently coexisted. Fig. 3 is a magnesium-potassium correlation plot of a special group

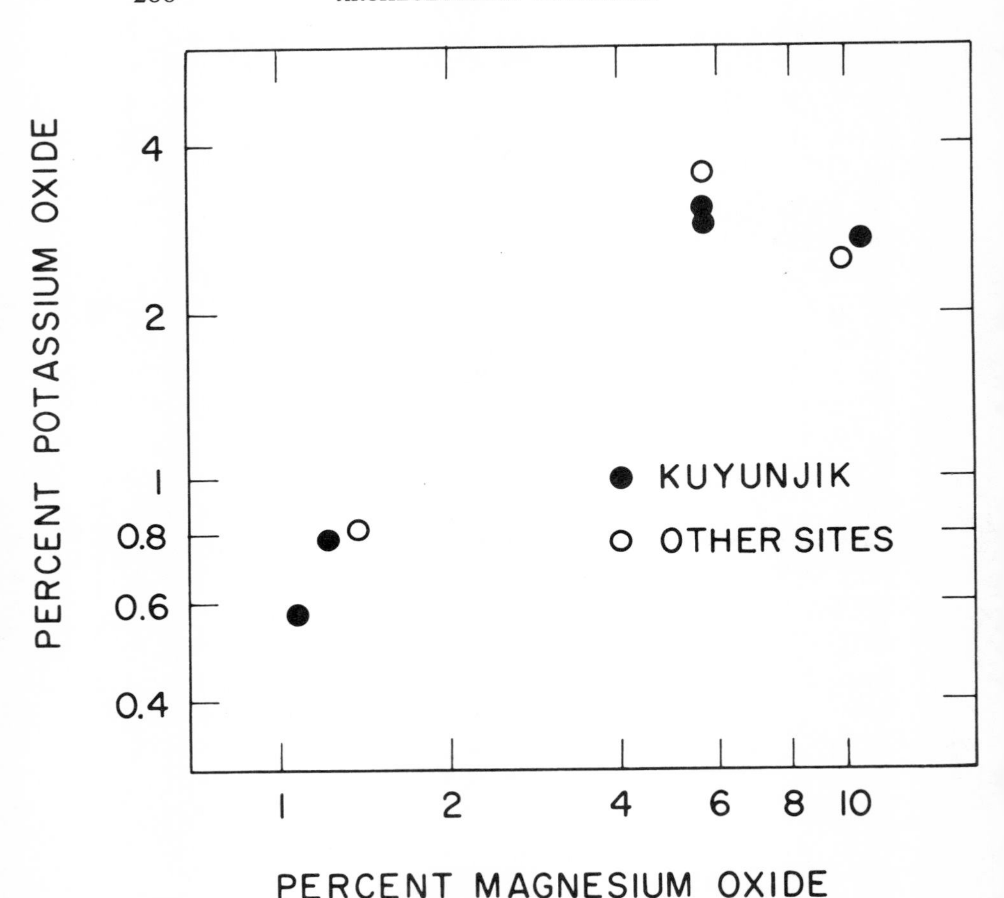


**Fig. 3. Occurrence of potassium and magnesium oxides in glasses of early Roman date from the Kuyunjik mound at Nineveh and other Mesopotamian sites.**

of specimens supplied to us by the British Museum. These specimens had all been excavated at Mesopotamian sites, six of them specifically coming from the Kuyunjik mound at Nineveh. Most of the specimens, including samples of the high magnesium type, are considered to be of Roman date.

The occurrence pattern of the two glass types shows (1) that each type of glass was produced or used for long periods of time in many widely separated geographic areas and (2) that in some areas both types of glass have occurred. The basic difference in composition thus does not appear to reflect local variation in the composition of specific raw materials. Rather the distinction more probably derives from the use of inherently different raw materials. Another observation to be made is that the correlation between high-magnesium and high-potassium content seems sufficiently pronounced to indicate the high concentration of both must come from the same component. The high-magnesium specimens were produced over a time span of at least 2500 years and in many different geographic areas. If the higher magnesium and higher potassium concentrations were contributed by separate raw materials it seems most unlikely that for this long period these materials always would have been used together. If the two elements were derived from separate raw materials and if a group of high-magnesium glasses were produced with the same potassium determining components as the low-magnesium glasses, the expectation would have been that more than three out of four of them would have had potassium concentrations falling statistically significantly below the normal range for high-magnesium glasses. It is clear that this does not happen because the potassium distribution curve for high-magnesium glasses in Fig. 1 shows no statistically unexpected tail extending into this low range. Despite the fact that the two potassium distributions overlap more than the magnesium distributions do they each unto themselves conform reasonably to normal expectancy for individual groups. If one concludes that but one component was fundamentally different in the two glass types then the correlation between magnesium and the alkali metals and the lack of correlation between magnesium and calcium both argue against the hypothesis that this component was limestone, i.e., that dolomite was exchanged for calcite. There are, moreover, alternate types of alkali known to have been used in antiquity which could explain the observed differences.

The low concentrations of magnesium and potassium in the low magnesium glasses, of the order of one percent magnesium oxide and a little over a half of a percent potassium oxide, are commensurate with the concentrations usually encountered in sand. The presence of feldspars and clays in sands normally introduces potassium equivalent to about a percent of potassium oxide. The amount of magnesium in sands is rather variable but concentrations of from 1 to 2 percent are not unusual. Therefore sand might have been the chief source of the potassium and magnesium normally found in low-magnesium glass and an alkali containing little of these elements might have been used in this type of glass. Natron, the naturally occurring sodium carbonate that is deposited through annual flooding and evaporation of certain Egyptian lakes, is such an alkali. Also Pliny [5] refers to the use of *nitrum* in glass. The glass manufacturing with which he would have been familiar would have utilized the low magnesium formulation. Potassium and magnesium concentrations have been neglected in previous analyses of natron, therefore we have analyzed for minor components the contents of two bags of natron taken from Eleventh and Eighteenth Dynasty tombs, which were supplied to us through the courtesy of the Metropolitan Museum of Art. The analyses were performed by the same spectrographic procedure as were the glasses. Because calibration was not carried out specifically with a sodium carbonate matrix the accuracy of results would be less than that obtained for glass specimens and the data are accordingly reported to only one significant figure. Potassium and magnesium oxides were present in these natron specimens only to the order of 0.1 percent, a low concentration consistent with the relatively small concentration of these oxides in low magnesium glasses. The observation that natron could have been the alkali in this type of glass, of course, is not submitted as sufficient proof that it was indeed so used.

Parodi [6] concluded from his analyses of early Egyptian glasses that they contained a distinctly higher concentration of potassium than would be anticipated had natron constituted their alkali component. The specimens for which he made separate potassium determinations were all of the high-magnesium type. In this conclusion the concentration of potassium contributed by sand was apparently neglected. Parodi did not report potassium in his analyses of sands. However, on the basis of all of our present information one would still tend to agree that there is more potassium in the high magnesium glasses than would be expected had they been formulated with natron. There are, however, at least two other sources of alkali that although they are primarily salts of sodium could have supplied the additional quantities of magnesium and potassium in the high-magnesium glasses. These alkalis are the residues obtained by evaporation of river waters and the ashes of certain plants from desert or sea coast regions. Turner [7] and others have discussed these possibilities extensively.

TABLE I.

Minor Constituents in New Kingdom Natron

| *Specimen No.* | *692* | *693* |
|---|---|---|
| *Description* | *From 11th Dynasty Tomb of Neket Re* | *From Embalming pits of Tutankhamen Tomb, 18th Dynasty* |
| *Constituent* | *Percent Concentration* | |
| *Lithium* ($Li_2O$) | 0.001 | 0.001 |
| *Potassium* ($K_2O$) | 0.07 | 0.2 |
| *Rubidium* ($Rb_2O$) | 0.0006 | 0.001 |
| *Magnesium* ($MgO$) | 0.03 | 0.2 |
| *Calcium* ($CaO$) | 0.3 | 1 |
| *Strontium* ($SrO$) | 0.04 | 0.03 |
| *Barium* ($BaO$) | 0 ($<0.0006$) | 0.001 |
| *Boron* ($B_2O_3$) | 0 ($<0.02$) | 0.02 |
| *Aluminum* ($Al_2O_3$) | 0 ($<0.1$) | 0.1 |
| *Phosphorous* ($P_2O_5$) | 0.01 | 0.03 |
| *Titanium* ($TiO_2$) | 0.01 | 0.06 |
| *Vanadium* ($V_2O_5$) | 0.03 | 0.03 |
| *Chromium* ($Cr_2O_3$) | 0.002 | 0.0009 |
| *Manganese* ($MnO$) | 0 ($<0.006$) | 0.01 |
| *Iron* ($Fe_2O_3$) | 0 ($<0.06$) | 0.1 |
| *Copper* ($CuO$) | 0 ($<0.0003$) | 0.001 |

Zinc, silver, tin, antimony, arsenic, lead, bismuth, zirconium, cobalt and nickel were all below sensitivity limits.

Both Pliny [5] and Agricola [8] refer to the evaporation of Nile River water to produce a salt residue of commercial importance. That residues so obtained from rivers of North Africa and the Middle East could contain the relative concentrations of sodium, potassium and magnesium found in high-magnesium glass is demonstrated in Table II. The mean concentrations of these elements in this type of glass are compared to those found in residues obtained by evaporating water from the Nile and Jordan rivers. That the concentrations of calcium in these residues are also roughly in proportion to the calcium in the glass is probably fortuitous. In a large scale evaporation of these waters, the more insoluble calcium salts are likely to separate out early and so be removed. It would seem more likely that calcium was introduced into both the high- and low-magnesium glasses chiefly by the same materials.

TABLE II.

Residues Obtained upon Evaporation of Water from the Nile and Jordan Rivers

| | *Nile** | *Jordan†* | *Mean of High Magnesia Glasses* |
|---|---|---|---|
| | *Percent Concentration of Seperate Ions* | | |
| *Sodium* | 13.14 | 18.11 | 9.9 |
| *Potassium* | 3.26 | 1.14 | 1.65 |
| *Magnesium* | 7.39 | 4.88 | 2.8 |
| *Calcium* | 13.31 | 10.67 | 4.8 |
| *Carbonate* | 36.02 | 13.11 | |
| *Sulfate* | 3.93 | 7.22 | |
| *Chloride* | 2.83 | 41.47 | |
| *Phosphate* | 0.59 | — | |
| *Silica* | 16.88 | 1.95 | |
| *Iron Oxide* | 2.65 | 1.45‡ | |

* From the Nile about two hours journey below Cairo. Analysis by O. Popp, Liebig's Annalen, 155 (1870), p. 344.
† From the Jordan near Jericho, Analysis by R. Sachsse, Inorg. Diss., Erlangen 1896. Both sets of data are quoted from F. W. Clarke, The Data of Geochemistry, U. S. Geological Survey Bulletin 770 (1924).
‡ Aluminum oxide included.

Plant ashes have long been considered as a possible source of alkali for ancient sodium rich glasses. Vegetation from deserts, the coasts or seas might be expected to have a sufficient predominance of sodium over potassium for them to have been used in these sodium rich glasses. However, at least moderate concentrations of both potassium and magnesium are to be expected in all plant ashes.

Hence if they were used in glass just before the tenth century it is much more likely that they would have produced a glass of the high- than of the low-magnesium composition. Some indication that plant ashes were used in the high-magnesium glasses may have been provided by our analysis of a Venetian goblet of the late Renaissance with typical high-magnesium composition. Both Biringuccio [9] and Neri,[10] writing in the sixeenth and seventeenth centuries respectively, describe the use of plant ash imported from Eastern Mediterranean countries for Venetian glass. Of course a number of additional analyses of Venetian glass of the Renaissance will be required to establish the degree to which the specimen already investigated may be regarded as typical.

Each of these proposed alkalis would be expected to contain troublesome concentrations of sulfates and chlorides which would have to be removed, at least in part, during processing of the batch materials. Turner has discussed in detail how such interfering ions might have been removed during the fritting process. One procedure by which sulfates could have been eliminated would be through the reduction of the sulfates into readily decomposible sulfites. Such reduction would be promoted by the addition of carbonaceous material to the glass batch. In this respect we were much interested to discover among fragments from an Egyptian glass factory at Lisht in a collection of such objects at the Metropolitan Museum of Art a clump of partially fused glass like material in which a lump of charcoal was imbedded. The identities of these components were confirmed by analysis. Although the glassy material encompassing the charcoal did not contain the constituent proportions of a completed glass it was composed of materials normally found in glass. This direct indication that charcoal might have been mixed with the glass forming materials at early stages of firing lends increased credence to the hypothesis that the elimination of sulfates through reduction is an ancient technique.

## OCCURRENCE OF MANGANESE AND ANTIMONY AS DELIBERATE ADDITIVES

Both manganese and antimony like magnesium occur in two distinct ranges of concentration in these ancient glasses. The distribution of the concentrations of these elements within our specimens are shown in Fig. 4. The separation into groups of concentration greater or less than roughly 0.1 percent is apparent.*

The details of the occurrence of these elements throughout our specimens have been presented in a recent paper [11] and will only be summarized here. The appearance of high concentrations of these elements was sufficiently independent of special geographic regions and glass types and sufficiently correlated with desirable properties in the glasses containing them to lead to the conclusion that in most

* A brief explanation of the vertical lines in Fig. 4 marked "lower limits of reliable analysis" is required. These lines indicate the average low sensitivity limits of the spectrographic analyses as carried out. Unfortunately the most sensitive antimony line appearing on our plates was not sufficiently intense to reveal the concentrations in all specimens. There were a number of instances of duplicate determinations in which the antimony line was just measurable in but one of the two determinations of the same specimen. Also specimens in which antimony was below detectable limits were frequently accompanied by closely related specimens in which a small concentration of antimony could be measured. Instances of this type have led us to believe that when antimony was not detected its concentration lay not greatly below the sensitivity limit. Therefore, when not detected antimony was estimated to be a concentration of one half of the detectable limit. Such estimations have led to the highly approximate distribution below the limit of reliable analysis.

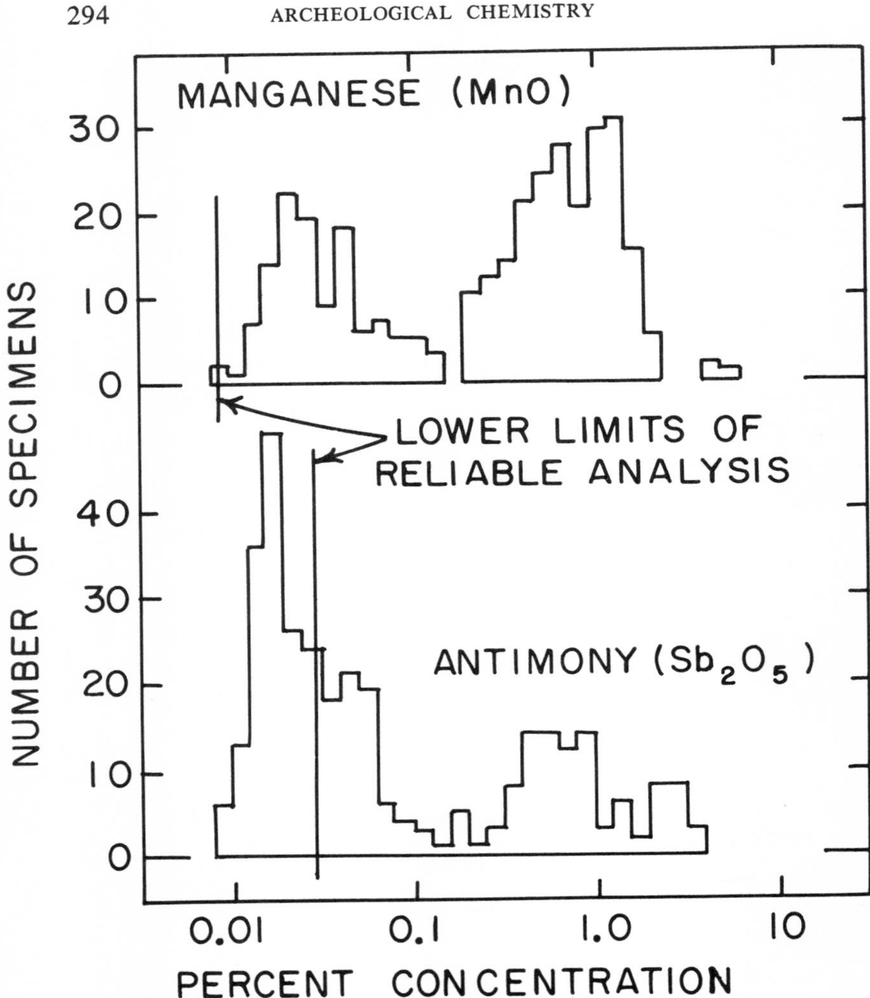


**Fig. 4. Occurrence of manganese and antimony oxides in ancient "soda-lime" glasses.**

instances they resulted from deliberate addition of manganese or antimony compounds to the glass batches.

In the earliest glasses, these two elements were used for quite separate purposes, but in later glasses they were utilized for a related purpose and sometimes even used together. From earliest times, manganese has been used as a colorant in characteristic reddish violet glasses. It also has been found in combination with cobalt, perhaps unintentionally as a result of natural association in ores, in some dark blue glasses. Compounds of antimony have made up the separated opacifying phase of some ancient glasses. Recently Turner and Rooksby [12] have undertaken an extensive study by means of X-ray diffraction of opacifying agents in ancient glass. They found calcium antimonates to be the opacifying agents in some white opaque glasses and lead antimonates in yellow opaque glasses. The white calcium and yellow lead antimonates when present in a blue matrix are responsible for some blue and green opaque glasses, respectively. The earliest specimen in which Turner and Rooksby found a compound of tin instead of one of antimony as an opacifying agent was dated to shortly before the middle of the fourteenth century A.D. They report however that, subsequent to this date, tin oxide was the chief opacifying agent encountered in white glasses.

Brill and Moll [13] have recently analyzed individual crystals of opacifying agents in ancient glasses by means of the electron microprobe. Their results have been in substantial agreement with Turner and Rooksby's.

We too have found considerable confirmation of these observations. Table III lists twenty one glasses dating from the nineteenth century B.C. to the fourteenth century A.D. in which antimony compounds are primarily responsible for opacity. Since most books currently dealing with the early technology of glass express the now questionable opinion that tin oxide was the opacifying agent in glasses of this type from these centuries it is important to discuss the instances in which tin has been found. The only specimens we have encountered in which tin oxide clearly has been used as an opacifier are some of Islamic date which will be discussed in a later section of this paper and some color band beads from the Amlash region. Unfortunately these beads could not be dated from the circumstances of their find. They were analyzed nevertheless because their style of execution was highly characteristic of the early Roman Imperial period. They are dark amber beads with flowing opaque white striations.

Therefore, like Turner and Rooksby, we can state that we have encountered many examples of the use of antimony compounds as opacifying agents in glasses from our earliest periods down through the Imperial Roman centuries. Also, like them, the earliest example of opacification by means of tin we have encountered, with the possible exception of the Amlash beads, appears to belong to the end of this period.

Contrary to these observations are a series of analyses by Neumann and his co-workers in which opacity was associated with a presence of tin. They reported tin in some second century A.D. Roman glass from Salona,[14] some Egyptian glass from the island of Elephantine [15] attributed to the Second-First Century B.C., some New Kingdom Egyptian glass from Tell el Armarna [16] and Gorub Medined,[17] and Babylonian-Assyrian glass from Nippur.[18] All of these glasses were either opaque or contained opaque inclosures which were included with the material analyzed. In none of these analyses was an appreciable concentration of antimony reported, nor in any of them in which the opaque phase was described as yellow was an appreciable concentration of lead reported. The contrast between these observations is most pronounced in the glasses of the second and early first millennia B.C. Current researchers have now found more than twenty specimens from this period which were opacified by compounds of antimony without once having encountered a specimen of this period opacified with a compound of tin. Neumann *et. al.* reported six such early glasses opacified with tin without finding one containing antimony. Also the only opacifying agents with a pronounced yellow color encountered in current investigations have been compounds of lead, lead antimonates in early glasses and lead stannates in later glasses. Three of the Neumann specimens from Gorub Medined included in the analyses were described as containing yellow inlayed decorations, but the specimens were specifically reported as containing no lead. It is to be noted that all six of these very early Neumann specimens were strongly colored by compounds of copper. We have encountered several examples in which copper in glass has been accompanied by tin and sometimes lead in proportions similar to those in which they are found in bronze. These occurrences will be discussed in a later section of this paper. There are also copper ores that contain considerable tin. Therefore, the addition of copper-containing substances to glass might accidentally have introduced tin. However, even though the presence of tin in the Neumann specimens might be explainable in such terms the absence of antimony and lead still would remain a problem. It might be noted that the chemical behaviors of tin, antimony and lead are somewhat alike and that the Neumann *et. al.* analyses were performed by wet chemical means. Unless the analyst were much on guard against it antimony and perhaps lead might have become included in a determination for tin or even

**TABLE III.**
**Concentrations of Antimony, Lead and Tin in Some Opaque Ancient Glasses**

| *Specimen No.* | Percent Concentration | | | *Description of Specimens* |
|---|---|---|---|---|
| | $Sb_2O_5$ | $PbO$ | $SnO_2$ | |
| *White* | | | | |
| 555 (1) | 2.3 | 0.078 | 0.0026 | White thread on blue sand-core vessel. Palace of Amenhotep III (14th century B.C.) |
| 224* | 1.98 | 0.0081 | 0.064 | Mycenaean ornaments (about 14th century B.C.) (224*, 225*) |
| 225* | 0.97 | 0.022 | 0.006 | |
| 847 | 3.4 | 0.023 | $<0.0021$ | Inlay strip from Palace of Lucius Verus (2nd century A.D.) |
| *Yellow* | | | | |
| 554 (1) | 0.93 | 3.2 | 0.0022 | Yellow threads on blue sand-core vessels. Palace of Amenhotep III (14th century B.C.) (554 (1), 555 (2)) |
| 555 (2) | 0.50 | 3.6 | 0.0048 | |
| 818 (1) | 0.55 | 2.5 | 0.055 | Yellow thread on blue sand-core vessel from Lisht (late second millennium B.C.) |
| 846 | 1.28 | 12.2 | 0.13 | Inlay strip from Palace of Lucius Verus (2nd century A.D.) |
| *Blue* | | | | |
| 760 | 2.5 | 0.0129 | 0.21 | Sand-core vessel from Tomb of Tuthmosis III (15th century B.C.) |
| 554 (0) | 0.42 | 0.0034 | $<0.0025$ | Sand-core vessel from Palace of Amenhotep III (14th century B.C.) |

| | | | | |
|---|---|---|---|---|
| 790* | 2.7 | 0.013 | <0.0024 | Mesopotamian cylinder seals (Late 2nd millenium B.C.) |
| 791* | 0.78 | 0.0027 | 0.045 | |
| 792 | 2.2 | 0.066 | <0.0023 | |
| 820 | 2.6 | 0.095 | 0.31 | Inlay datable to reign of Ptolemy V (2nd century B.C.) |
| 913 | 0.94 | 0.11 | 0.0022 | Mosaic cube from Grotto of Tiberius (1st century A.D.) |
| 690 (1) | 2.7 | 0.023 | 0.010 | Upper layer of Roman cameo and of over-coated relief plaque (1st century A.D.) |
| 691 (1) | 2.2 | 0.059 | 0.021 | |
| 23 | 2.2 | 0.031 | 0.0027 | Blown vessel from Begram (1st to 3rd century A.D.) |
| 724 | 0.75 | 0.59 | 0.12 | Plate excavated at Athens agora (2nd to 4th century A.D.) |
| 832 | 1.90 | 0.17 | 0.0047 | Round slab from Villa Leutersdorf (2nd to 4th century A.D.) |

* Badily hydrolyzed.

have been mistaken for tin. Unfortunately the specific methods used for these analyses were not described. It is to be hoped that the uncertainties introduced by these highly conflicting results will be more fully resolved as continued studies are forthcoming. It would be particularly helpful if some of Neumann's specimens could be obtained for re-examination.

Opaque glasses have not been the only ones containing high concentration of antimony. In clear, colorless glasses of the latter half of the first millennium B.C. from Olympia, Gordion and Persepolis, and in a large number of Roman glasses, we have regularly encountered a concentration of antimony oxides as great as in opaque glasses. The difference in the properties of antimony in the opaque and clear glasses seems to reside in its state of chemical valence. The antimony is in its higher, pentavalent state in the opaque glasses and in a lower state of valence, probably trivalent, in the clear glasses.[1] These glasses are particularly free of the discoloration one might expect to arise from the quantity of iron they contain. This is not too surprising because some compounds of antimony should be capable of oxidizing and reacting to form chemical complexes with the iron in the glass. These processes would tend to remove the typical "bottle green" discoloration arising from reduced iron. Also the concentration of antimony present was sufficiently great to have reacted with the concentration of iron within the glass. It would seem reasonable that some glassmakers of this period when attempting to make opaque white glass altered conditions sufficiently so that the antimony was reduced and the process for making this exceptionally colorless limpid glass was discovered.

Colorless clear glasses from the Near East of the latter half of the first millennium B.C. regularly continued to have antimony until about the late first century B.C. when one encounters colorless glasses in which manganese appears to have replaced antimony. Since some compounds of manganese will serve like compounds of antimony to oxidize divalent iron in glass and thereby decolorize the glass, the exchange might well indicate a simple replacement of one decolorant by another. Pliny wrote of the addition of *magnes lapis* to glass batches. This "magnet stone" has been variously interpreted, 1) as magnesia or a magnesium containing limestone, 2) as a magnetic oxide of iron or, 3) as an ore or compound of manganese. However, Roman glass and the occidental glass that preceded it for several centuries was of the low magnesium type, containing relatively a minimum concentration of magnesium, and it is thus indicated that magnesium as such was not an additive. Also, early Roman glass contains no more iron on the average than other ancient glasses nor would any benefit have been derived from additional iron in the batch. Further our data would indicate that manganese was frequently added intentionally to glass in the first century A.D. It would seem quite possible that Pliny's reference was to this addition.

Glasses with high antimony, with high manganese, or with both together, were produced in the Middle East and Western Asia until roughly the latter part of the fourteenth century A.D. From this time on we have found manganese alone in decolorized glasses. In Italy and Southern Germany both of these two elements also were used widely as decolorants during the Imperial Centuries. However, in Western Europe the use of antimony died out even earlier than in the Eastern Mediterranean region. In clear glasses from European tombs which can be dated to later than the fourteenth century we have encountered only low concentrations of antimony, but usually manganese continues to appear in such specimens. It seems a most significant coincidence that Turner and Rooksby have found tin to have largely replaced antimony in opaque glasses at this same time.

Eighty-four percent of our specimens have contained manganese, antimony or both in high concentrations that seem to indicate deliberate addition. One concludes therefore that compounds of these elements were widely used by glass manufacturers in antiquity.

## OCCURRENCE OF LEAD AS AN INTENTIONAL ADDITIVE

The occurrence of lead in yellow opacifying agents has been discussed. It is significant to note, though, that decorative inlays of opaque white glass, containing calcium antimonates, and of opaque yellow glass, containing lead antimonates, frequently occurred together on New Kingdom sand core vessels. The makers of these vessels must also have noticed that if they tried to prepare an opaque glass with lead without antimony the glass would remain clear. These striking differences in the behavior of compounds of antimony and of lead must surely have impressed these early glassmakers and provided for them a means of differentiating between antimony and lead.

There were some types of opaque glass, however, to which lead was customarily added without antimony. These are the opaque red glasses which owe their opacity to dispersed cuprous oxide and metallic copper, and some Islamic opaque white glasses in which tin oxide is the separated phase. Partial analyses of fifteen opaque red glasses are listed in Table IV. It will be noted that the large majority of these specimens contain a substantial concentration of lead. The constituents of these glasses, other than lead, copper, and in some cases iron, are in roughly the same relative proportions as in normal glasses of their periods. That is to say, it would appear that the lead, copper and sometimes iron had been added to a normal basic glass.

A somewhat analogous situation occurs in the specimens listed in Table V. These specimens are three opaque, creamy-white coin weights and an opaque green mosaic tessera. All are of early Islamic date. As in the opaque red glasses the glass matrix to which the opacifying agent and lead were added was the normal glass of its period. Turner and Rooksby have examined the three coin weights by X-ray diffraction and have identified the separated phase within them as tin dioxide. Thus the lead does not constitute part of the separated opacifying phases in these glasses. Why then was it added? The answer may lie in the procedure by which uniform opaque glasses are usually produced. Best results are obtained if the opacifying agent is soluble in the glass melt at high temperatures. It then "strikes," that is, precipitates out, when the glass is held for a period of time at an intermediate temperature. In this way many small uniform crystals of the opacifying phase are formed. The presence of lead oxides should add materially to the solubility of both tin and copper oxides in the glasses and hence assist in the process just described.* Matson [20] has stated that the presence of some lead would be essential in the manufacture of the opaque red glasses.

Two of the opaque red glasses in which the lead concen-

* Dr. Robert H. Brill of the Corning Museum of Glass who independently has arrived at the above conclusion as a probable explanation of the presence of lead in the opaque red glasses has informed us that the Corning Glass Laboratories have found that the presence of lead oxides will increase the solubility of copper and tin oxides in glass.

TABLE IV.

Partial Analyses of Some Ancient Opaque Red Glasses

| | *Percent Concentration* Lead PbO | Copper $Cu_2O$ | Iron FeO | *Description of Specimens* | *Analyst* | |
|---|---|---|---|---|---|---|
| | nil | 10.28 | 8.24 | Indian from Assam —not dated | Ullah | (a) |
| | nil | 0.49 | 7.01 | Indian from Nalanda —probably 5th cent. A.D. | Ullah | |
| *(antimony = | 0.23<br>2.0) | 11 | 1.02 | Egyptian—late 7th or early 6th cent. B.C. | Sayre No. 819 | |
| (antimony = | 4.6<br>0.63) | 0.71 | 4.7 | Found in Roman grave at St. Aldegund in the Rhine region—probably 1st half of the 4th cent. A.D. | Sayre No. 18 | |
| | 1.28<br>3.02<br>6.28 | 2.09<br>2.52<br>4.40 | 0.77<br>1.41<br>1.17 | Egyptian from island of Elephantine—about 1st century B.C. | Kotyga | (b) |
| | 15.51 | 11.03 | 2.10 | From Pompeii—1st cent. A.D. | Pettenkofer | (c) |
| (antimony = | 15.9<br>0.33) | 1.74 | 1.66 | Roman from palace of Lucius B.C. | Sayre No. 848 | |
| (antimony = | 21<br>0.42 | 7.1 | 0.89 | Egyptian—4th cent. B.C. | Sayre No. 809 | |
| (antimony = | 22.8<br>4.07 | 13.58 | — | Nimrud—8th–6th cents. B.C. | Turner | (d) |

| | | | | | | |
|---|---|---|---|---|---|---|
| | 31.10 | 4.67 | — | “Roman Egyptian” | Jackson | (e) |
| | 32.85 | 9.86 | — | Roman glass found at Tara Hill in Ireland | Reynold | (f) |
| | 35.7 | 13.1 | 1.6 | Egyptian from Canopus of Roman date | Mercanton | (g) |
| | 38.93 | 5.98 | — | Indian from Taxila of Roman date | Ullah | (a) |

* antimony as percent $Sb_2O_5$.

a) Varshney, Y. D., “Glass in Ancient India,” *Glass Ind.,* 31, (1950), pp. 632–634.
b) Neuman, B. and Kotyga, G., “Antike Gläser, ihre Zusammensetzung und Färbung,” *Z. Ange. Chemie* 38 (1925), pp. 857–64.
c) Pettenkofer, M., “Ueber einen antiken rothen, Glasflusz (Hamatinon) und ueber der Aventurin-Glas,” *Dingler's Polytechnisches Journal,* 145 (1857), pp. 122–134.
d) Turner, W. E. S., “Glass fragments from Nimrud of the Eighth to Sixth Century B.C., *Iraq,* 17 (1955). pp. 57–68.
e) Collie, J. N., “Notes on the 'Sang de Boeuf' and the Copper-red Chinese glazes,” *Trans. Ceram. Soc.,* 17 (1917–18), pp. 379–384.
f) Ball, V., “On a Block of Red Glass Enamel Said to Have Been Found at Tara Hill,” *Trans. Roy. Irish Acad.,* 30 (1893), pp. 277–281.
g) Minutoli, H., *Anfertigung und Nutzanwendung der farbigen Gläser bei den Alten* (Berlin, 1836), p. 34.

tration was relatively low contained sizeable concentrations of antimony which should behave similarly to lead in affecting solubility within the glasses. Also three opaque red glasses with low lead contained high concentration of iron oxide (5 to 8 percent). One might think the iron oxide might have come out of solution in these glasses contributing along with copper to the opacity. However, we tested the one glass of this type we encountered, specimen No. 18, by X-ray diffraction and observed only copper and cuprous oxide in its diffraction pattern. It would seem probable, although it has yet to be established, that the high concentrations of iron oxides in these glasses would also facilitate the solution of copper oxides at higher temperatures, and that it too might have been added for this purpose. It may, however, have been added only to deepen the color of the glass.

TABLE V.

Tin and Lead Concentrations in Some Islamic Opaque Glasses

| *Percent Concentration* | | *Specimen* | |
|---|---|---|---|
| *Tin* ($SnO_2$) | *Lead* ($PbO$) | *number* | *Description* |
| 7.9 | 26 | 278 | Late 'Abbāsid coin weight, al-Mustadi* |
| 5.5 | 28 | 295 | 'Abbāsid coin weight, al-Nasir* |
| 7.2 | 23 | 300 | Fatimid coin weight, al-Hafiz* |
| 2.2 | 43 | 777 | Mosiac cube from Umayyad mosque |

* Names of officials appearing in inscriptions on the weights.

In the opaque glasses just described a specific use of lead in glass is encountered for which antimony probably could have been substituted. With but few exceptions, however, antimony was found in much lesser quantities than lead. In the same context of period and region in which many of the opaque red glasses were found, clear, colorless glasses with significant concentrations of antimony but relatively minor amounts of lead were likewise found. We have now analyzed some 230 ancient glasses that are relatively clear and free of colors of which nearly 70 contained relatively high concentrations of antimony without having once encountered a clear, colorless specimen containing lead, or for that matter tin, in concentration greater than would have been expected to arise from normal minor impurities in the other glass constituents. We know from modern lead crystal that lead could perfectly well have been accommodated in these glasses. If, as often has been maintained, ancient artisans were confused about the separate identity of lead and antimony to the extent that they would mistakenly use one for the other, would not one have expected much greater overlap in their occurrence in ancient glass?

## OXIDIZED BRONZE AS AN ADDITIVE

In discussing his analyses of Rhenish-Roman bronzes Geilmann observed that in several of his glasses copper, tin and lead were present in proportions similar to those in which they occur in bronze. An oxidized or corroded bronze might well have been included in the glass melts to introduce copper as a colorant. There have been a number of similar occurrences among our data. In Table VI are listed nine of our specimens in which copper, tin and lead parallels the composition of bronze for the period of their manufacture.

Table VI.

Some Ancient Glasses Containing Copper, Tin, and Lead in Proportions Characteristic of Bronze

| Percent Concentration | | | Specimen number | Description |
|---|---|---|---|---|
| Copper (CuO | Tin $SnO_2$ | Lead PbO in Glass) | | |
| | | | *Second Millennium B.C.* | |
| 1.45 | 0.21 | 0.013* | 760 | Sand-core vessel from tomb of Tutmosis III |
| 0.79 | 0.083 | 0.005* | 971 | Heavy glaze from walls of Elamite Ziggurat at Zchoga Zambil |
| 0.39 | 0.064 | 0.008* | 224 | Ornament from Mycenae (partially hydrolysed) |
| 85.4 | 8.5 | 0.1† | | Average of 14 analyses of early Egyptian bronzes tabulated by Lucas |
| | | | *First Millennium B.C.* | |
| 2.8 | 0.31 | 0.09 | 820 | Inlay datable to Ptolemy V |
| | | | *Early First Millennium A.D.* | |
| 0.69 | 0.12 | 0.59 | 724 | Plate from the Athens Agora |
| 0.75 | 0.24 | 0.16 | 593 | Roman bowl with embedded millefiori |
| 2.1 | 0.11 | 0.09 | 743 | Pitcher from Valerian tomb beneath St. Peters in Rome |
| 2.7 | 0.17 | 0.13 | 835 | Diatretum fragment from the Moselle Valley |
| 2.0 | 0.29 | 0.88 | 845 | Millefiori fragment from palace of Lucius Verus in Rome |
| 74.0 | 9.5 | 15.0 | | Average of 6 Rhenish-Roman bronzes analyzed by Geilmann |

* Approximately equal to mean lead concentrations for all glass of this period.
† One bronze of exceptionally high lead concentration (8.5%) not included in this average.

It is well known that the earliest Western bronzes as a rule contained little lead. Lead began to be introduced with greater frequency in the classic Greek bronzes and is regularly encountered in Roman bronzes. This change is reflected in these glasses. Three specimens of second millennium B.C. glass are compared to the average of fourteen analyses of early Egyptian bronzes which were tabulated by Lucas. The glasses contain little more lead than other glasses of this period, and with but one exception (noted on the table) the bronzes contained only trace amounts of lead. Five specimens of Imperial Roman glass contained approximately as much lead as tin, paralleling the average shown in six analyses of Rhenish-Roman bronzes by Geilmann.

It is perhaps significant to note that within our specimens all instances of glass manufactured before or during the early Roman Imperial period that have contained a significant concentration of tin have either been of the type considered here, in which the tin is associated with about a ten to one excess of copper indicating bronze as a common source of both elements, or they have been of the high lead containing opaque red type in which the tin was likely introduced as an impurity in the lead. Considered by themselves, there is no direct evidence in our data that tin was directly and intentionally used in glass before the Roman Imperial period.

## ACKNOWLEDGMENTS

We are deeply grateful to the many individuals and institutions that have furnished us with glass specimens for this study. Their number is now so large that it would be very difficult to acknowledge them individually in a short paper. The main body of data from which the observations presented here are derived will soon be published in detail with proper descriptions of the objects analyzed and their sources.

Mr. Rutherford J. Gettins was of considerable help in the establishment of some of the methods of analysis used in this study. Mrs. Patricia Walsh has been of extensive assistance in carrying out the technical details of many of the analyses.

## REFERENCES

1. Geilmann, W., and Jenemann, H., "Die Bestimmung geringer Phosphatgehalte in Glasern," *Glastech. Ber.*, 26, (1953), pp. 341–346. Geilmann, W., and Bruckbauer, T., "Beitrage zur Kenntnis Alter Glaser II. Der Mangangehalt alter Glaser," *Glastech. Ber.*, 27 (1954) pp. 456–459. Geilmann, W., "Beiträge zur Kenntnis Alter Gläser III," die Chemische Zusammensetzung einiger alter Glaser, inbesondere deutscher Glaser des 10. bis 18. Jahrunderts, *Glastech. Ber.* 28 (1955) pp. 146–156.

2. Bezborodov, M. A., *Glass Manufacturing in Ancient Russia,* Akad. Nauk BSSR, (Minsk, 1956).

3. Beck, H. C., and Seligmann, C. G., "Barium in Ancient Glass," *Nature 133* (1934) p. 982.

Seligmann, C. G., Ritchie, P. D., and Beck, H. C., "Early Chinese Glass from pre-Han to T'ang Times," *Nature* 138 (1936) p. 721. Ritchie, P. D., "Chinese Glass from pre-Han to T'ang Times," *Technical Studies in the Field of Fine Arts,* 5 (1937) pp. 209–220. Seligmann, C. G., and Beck, H. C., "Far Eastern Glass: Some Western Origins," *Bulletin of the Museum of Far Eastern Antiquities,* No. 10, (Stockholm, 1938).

4. Smith, R. W., "Archaeological Evaluation of Analyses of Ancient Glass," in *Advances in Glass Technology,* F. R. Matson and G. E. Rindone (eds.) (New York, 1936), p. 283.

5. Pliny, *Natural History,* Book XXXVI, chaps. 65 and 66, and Book XXXI, chap. 46.

6. Parodi, H. D., "La verrerie en Egypte," thesis, Grenoble, 1908.

7. Turner, W. E. S., *Studies in Ancient Glasses and Glassmaking Processes,* Part V, "Raw Materials and Melting Processes." *Trans. Soc. Glass Tech.* 40 (1956) pp. 277–300.

8. Agricola, Georgius., *De Re Metallica,* Book XII, 1556.

9. Biringuccio, V., *Pirotechnia,* Book II, chap. 14, 1540.
10. Neri, Antonio., *L'Arte Vetraria,* 1612.
11. Sayre, E .V., "Intentional Use of Antimony and Manganese in Ancient Glasses," in *Advances in Glass Technology,* F. R. Matson and G. E. Rindone (eds.) (New York, 1963), p. 263.
12. Turner, W. E. S., and Rooksby, N. P. "A Study of the Opalescing Agents in Ancient Opal Glasses throughout Three Thousand Four Hundred Years," *Glastechnische Ber.* 32K (1959) pp. 17–28. Sixth International Congress on Glass, Part II, 1962.
13. Brill, R. H. and Moll, S., "The Electron Beam Probe Microanalysis of Ancient Glass," in *Advances in Glass Technology,* F. R. Matson and G. E. Rindone, (eds.) (New York, 1963), p. 293.
14. Neumann, B., "Antike Gläser II," *Z. Angew. Chemie* 40 (1927), pp. 963–982, Specimen Nos. 76 and 81.
15. Neumann, B., "Antike Gläser, ihre Zusammensetzung und Färbung," *Z. Angew. Chemie* 38 (1925), pp. 857–864, Specimen No. 17.
16. *Ibid.,* Specimen No. 9.
17. Neumann, B., "Antike Gläser IV," *Z. Angew. Chemie* 42 (1929), pp. 835–838, Specimen Nos. 95, 96, 97 and 98.
18. *Ibid.,* Specimen No. 100.
19. Sayre, *op. cit.*
20. Matson, F. R., Analyses of Various Substances from Persepolis, appendix in Schmidt, E. F., *Persepolis II,* (Chicago, 1957), p. 131.

# L'Analyse Critique au Service de l'Histoire des Métaux Anciens

Par Adrienne R. Weill

Le passé est une croyance . . .
Paul Valéry.

## Introduction

Qu'il s'agisse de conserver ou simplement de classer un objet ancien, l'expert ne peut plus actuellement se contenter d'une description extérieure des formes ou du décor. Pour établir la fiche d'identité d'une pièce en toute conscience, il faut un jugement parfaitement objectif. Ainsi s'impose le recours à l'analyse.

Aujourd'hui encore ce seul mot fait trembler. Il évoque ces désintégrations qui ont procuré des triomphes éclatants aux chimistes jusqu'au moment où sont apparues les méthodes physiques non-destructives.

Certaines d'entre elles—nous l'avons dit en dressant un bref inventaire de celles qui sont d'ores et déjà bien éprouvées—ne sont en fait que le prolongement de l'observation.[1] Nous n'y reviendrons pas aujourd'hui. Ce que nous nous proposons d'étudier c'est la signification des résultats ainsi obtenus tant du point de vue du métallurgiste que de celui de l'historien ou de l'archéologue.

## Choix des méthodes

Selon la nature de l'échantillon qui lui est soumis, de manière tout à fait générale, deux sortes de problèmes se présentent à l'analyste.

Dans une première catégorie, l'échantillon a été conçu et préparé pour les besoins de la cause, parfois même pour éprouver la sensibilité d'une technique particulière, compte tenu des phénomènes physiques ou chimiques mis en jeu. Cependant dans un laboratoire d'applications, l'échantillon est absolument indépendant de l'analyse, c'est-à-dire qu'il s'agit de déchiffrer un objet quelconque d'origine naturelle, ou résultant d'une fabrication plus ou moins bien connue: à l'analyste de faire de cet objet un échantillon valable, de trouver la méthode appropriée à son destin d'échantillon.

Dans le premier cas en effet le spécimen n'avait d'autre vocation que celle d'être soumis à l'analyse; dans le second, l'objet ne répondra qu'à des questions bien posées, compte tenu des restrictions inhérentes à chaque mode d'essai.

En particulier l'objet ancien—plus précisément un fragment ou une pièce en métal, seul cas que nous considérerons ici—entre au laboratoire avec un signalement totalement étranger aux techniques qui vont lui être appliquées. Sa provenance ne coïncide généralement pas avec son origine, elle est un épisode ou une étape de son histoire métallurgique. Deux langages sont employés simultanément pour le décrire, et il convient de les accorder. Par exemple le musée confie au laboratoire un fermoir de collier en électrum. Il s'agit, au laboratoire, de situer ce témoin sur le diagramme ternaire or-argent-cuivre, compte tenu de tous les traitements volontaires ou fortuits subis par ces éléments entre l'instant où ils ont été extraits de leurs minéraux respectifs et celui où l'objet est devenu échantillon.[2]

Il est normal de n'utiliser que des examens totalement non-destructifs pour ces pièces non reproductibles, c'est-à-dire que l'aspect de l'objet, tel que l'oeil peut le percevoir avec ou sans microscope doit demeurer inchangé (sauf en ce qui concerne le dégagement éventuel de produits de décomposition) après la suite des essais destinés à fournir une estimation de la composition chimique et de l'état structural.

Cette exigence eût été il y a seulement une quinzaine d'années extrêmement restrictive du point de vue des opérations courantes du laboratoire, que nous pourrions définir comme celles qui peuvent être menées avec les appareils d'observation et de mesure disponibles sur le marché commercial. Ainsi, à propos de l'électrum, nous avons montré que l'association des examens superficiels par diffraction de rayons X et d'une mesure globale, en l'occurrence celle du poids spécifique, suffisait pour reconnaître l'intention de l'artisan quant à la composition chimique de son oeuvre.

Mais ce n'est pas là une méthode générale, applicable à tous les objets. Par exemple, la seule présence d'un oxyde rend illusoire la déterminaticn du poids spécifique. Tout au plus, cette mesure, précieuse du fait de sa simplicité et des résultats qu'elle fournit à peu de frais, permet-elle de savoir par des comparaisons répétées, si telle monnaie est oxydée ou non, éventuellement si l'altération est superficielle ou a gagné la masse.

D'autres associations de techniques non-destructives

seraient-elles plus avantageuses ? C'est la question qu'il est permis de se poser et de soumettre à l'expérience.

Se refusant au massacre de l'échantillon, l'analyste écarte les méthodes chimiques. S'il ne peut obtenir de l'archéologue l'autorisation de polir une parcelle de surface—ou d'y laisser une trace d'étincelle—métallographie et spectrographie dans l'ultra-violet seront également bannies. Il reste donc principalement l'utilisation des rayons pénétrants, rayons X, éventuellement radiations de haute énergie fournies par les éléments radio-actifs, ou activation de l'échantillon, compte tenu des précautions indispensables en ce qui concerne chaque mode d'investigation.

En certains cas particuliers d'autres méthodes, comme celles faisant appel aux propriétés magnétiques, sont exploitables avec grand profit. Nous n'en parlerons pas ici, et nous bornerons notre propos aux seules techniques de rayons X.
Nous en rappellerons brièvement le principe, ne serait-ce que pour reprendre pleine conscience des phénomènes mis en jeu.

Cet exercice est indispensable avant d'aborder des problèmes comme ceux que risquent de présenter les métaux anciens, car plus une technique est courante, plus elle risque d'être figée par l'usage, si bien que ses limitations ou ses possibilités d'extension ne sont plus totalement présentes à l'esprit.

L'analyse d'un objet chargé d'inconnu exige un effort critique de l'opérateur tout autre que l'habileté exercée pour des contrôles routiniers—comme l'orientation d'un monocristal en diffraction ou l'analyse d'un échantillon homogène en fluorescence. Cette habileté est dépensée en vain, si elle n'est accompagnée d'un examen de conscience, d'un retour constant aux principes, et d'une enquête scrupuleuse de tout ce qui est sous-entendu ou négligé dans les opérations quotidiennes.

## Techniques d'analyse aux rayons X

Trois méthodes principales faisant appel aux propriétés physiques des rayons X sont aujourd'hui au service de l'analyse. La plus connue, de par ses usages médicaux, est la radiographie, uniquement fondée sur l'absorption différentielle des rayons X par les divers éléments chimiques, ou par les diverses épaisseurs d'une même combinaison dans un objet quelconque.

Ainsi, grâce à la radiographie, on peut effectuer, par exemple, par voie directe la mesure de la densité d'un spécimen homogène de forme convenable, comme une lame à faces lisses. Si l'hétérogénéité est constituée par un revêtement épais, celui-ci apparaîtra sur la radiographie, qui est à l'échelle de l'objet, c'est-à-dire macroscopique. Si l'une des faces de l'objet, ou même chacune d'elles, porte des gravures en creux ou des décors en relief, ceux-ci figurent sur la radiographie, même si la pièce n'est ni dégagée de son environnement [3], ni de la gangue qui la recouvre.[4] Il y a donc, par le seul jeu de l'absorption différentielle, des possibilités de prolongation de l'observation dans un domaine de transparence situé bien au-delà de celui du spectre visible. Pour les objets trop épais pour l'analyse par les radiations X, ou bien dans le cas où le procédé serait jugé plus propice—par exemple sur un champ de fouilles—les mêmes expériences de transparence, d'ombres chinoises en quelque sorte, peuvent être effectuées avec les radiations gamma émises par les substances radioactives artificielles.

Ce phénomène sélectif d'absorption joue encore un rôle prépondérant dans l'application des deux autres méthodes courantes de rayons X, la diffraction et la fluorescence. Il ne sera plus, comme en radiographie, le centre même des expériences, cependant il interviendra à chaque instant et commandera en quelque sorte le domaine d'application de toute méthode fondée sur l'interaction d'un rayonnement dit pénétrant et de la matière sur laquelle il est dirigé.

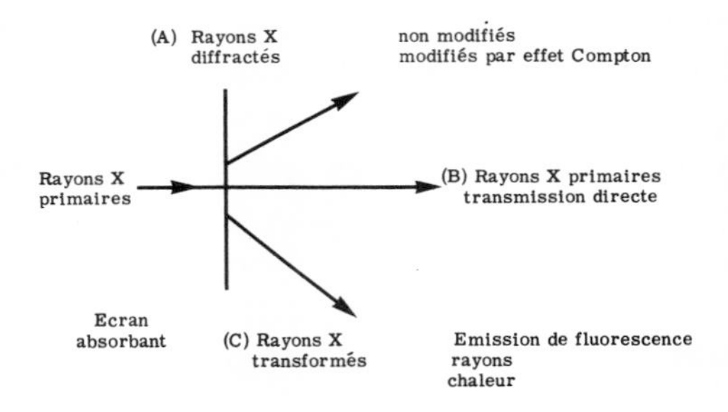


**Fig. 1. Passage des rayons X à travers la matière.**

Ceci est si vrai que dans l'un des ouvrages classiques sur "l'interprétation des diagrammes de diffraction de rayons X," [5] le schéma donné à propos de l'absorption résume en fait le principe des trois méthodes d'investigation de la matière par les rayons X. (Fig. 1 d'après la référence (5), p. 29).

Ce qui vient d'être dit au sujet de la radiographie con-

faussaire, contemporain ou non de la frappe originale, titre exact de la monnaie vraie, éventuellement recherche de l'origine des minerais utilisés, etc. . . .

Remarquons au passage que pour l'homme de laboratoire, comme pour l'historien, des problèmes fascinants se posent à propos des fausses monnaies de tout âge. Toute pièce de titre dégradé, toute monnaie fourrée ou saucée pose en effet des problèmes complexes, témoignant d'efforts techniques beaucoup plus ingénieux que la frappe d'une médaille en métaux précieux et apparemment inaltérables.

L'artisan d'autrefois qui a réussi à suppléer grâce à son habileté et à son ingéniosité à un défaut de matières premieres se rapproche ainsi à plus d'un titre de l'ingénieur contemporain, qui, dans des domaines très différents, n'excluant d'ailleurs pas ceux de la monnaie officielle, ne peut guère négliger le facteur économique dans la conception de ses oeuvres. C'est donc avec une certaine curiosité que nous aborderons l'examen de cette pièce argentée et ternie, provenant d'un trésor récemment exhumé dans le département de la Marne, où il avait été abandonné à l'époque des invasions romaines.

Déchiffrée au Cabinet des Médailles (Paris), cette pièce porte au droit l'inscription : IMP GORDIANUS PIUS FEL AUG, un buste radié à droite, vêtu du paludamentum vu de dos. Au revers, on lit : IOVI STATORI; on voit Jupiter debout tenant une haste et un foudre. La pièce a été frappée à Rome vers 240. On sait que l'empereur Gordien III, dit le Pieux, a régné de 238 à 244.

Le moyen bronze ne laisse pas apercevoir son décor: droit et revers sont envahis par les oxydes. L'inventeur, Monsieur Michel Fleury, Inspecteur Général des Fouilles de la Seine l'a trouvée dans les chantiers de la rue Pierre Nicolle (Paris. V[e]), auprès de la voie romaine qui porte

**Fig. 3. Pièce saucée à l'effigie de Gordien le Pieux; Rome (vers 240).**

aujourd'hui les noms de rue et faubourg St Jacques, parmi des tombes gallo-romaines datant du premier ou du second siècle après J. C.

Avant d'être photographiée (fig. 2), la pièce a été "nettoyée" par la prise successive sur le droit de deux répliques en vernis nitrocellulosique.[8] Cette opération dégage partiellement les oxydes, qui sont d'ailleurs ainsi conservés point par point sur l'empreinte, cependant l'effigie originale demeure très effacée, pratiquement indéchiffrable, malgré l'aspect métallique repris par la plus grande partie de la surface.

b) *Analyses rapides par fluorescence.* Les deux pièces ont été analysées par spectrométrie sous vide selon la technique courante : la radiation primaire (molybdène) est excitée sous 48 kV avec un débit de 24 mA. Un monocristal de fluorure de lithium, placé dans le vide comme l'échantillon et la fenêtre d'entrée du compteur à scintillations, renvoie sous différents angles, selon leurs propres longueurs d'onde, les radiations émises par les divers éléments constituant l'échantillon. Chaque pièce est irradiée sur son revers, la hauteur du faisceau est d'environ 1 cm, sa largeur de 2 cm. L'examen intéresse donc la majeure partie de la surface de la pièce d'aspect argenté.

Après amplification, la réponse du compteur est enregistrée sur papier. Les différentes radiations sont identifiées selon l'angle sous lequel elles apparaissent, déduction faite des inévitables raies d'émission primaire (molybdène).

Bien que l'importance des pics représentant les différents éléments soit en relation avec leur taux dans l'alliage, l'analyse demeure qualitative, tout au plus semi-quantitative, tant que l'on ne dispose pas d'étalons de comparaison. Ces étalons sont idéalement constitués par des séries d'alliages artificiels de compositions bien connues et se distinguant entre elles par la variation systématique du taux d'un seul constituant. Encore une fois, pour les échantillons complexes, ceci est dû aux phénomènes d'absorption.

Néanmoins l'analyse rapide peut renseigner sur l'intention de l'auteur de l'oeuvre, dans la mesure où son interprétation permet de distinguer une addition non fortuite d'une impureté de la mine, ou d'une trace introduite par le processus d'élaboration.

C'est dans cet esprit que nous avons classé les résultats des analyses par fluorescence, donnant pour la monnaie gordienne une estimation très grossière des proportions apparentes (Tableau), et signalant celles des traces que l'on s'attendait à trouver et qui sont absentes de l'enregistrement —fer et antimoine par exemple. Le cas de l'or n'a pas été étudié à fond: un très faible pic pourrait indiquer la présence de ce métal soit à l'état d'impureté de la mine d'argent, soit par contamination. Le trésor contenait également des pièces d'or, une diffusion aurait pu se produire au cours des siècles si les surfaces d'or et d'argent avaient été en contact, surtout si elles avaient été serrées les unes contre les autres.

c) *Diffraction des rayons X*. L'enregistrement par compteur est nettement avantageux pour la fluorescence, car tout document qualitatif peut à un moment quelconque servir de base à un dosage quantitatif à partir d'une gamme appropriée d'étalons artificiels.

Il en va tout autrement pour la diffraction : le compteur intègre des détails de texture que la photographie conserve. De plus l'enregistrement du compteur s'adresse à des plages d'un à deux centimètres carrés, tandis que, par photographie, il est possible de procéder à des sondages locaux limités à environ un demi millimètre carré. Néanmoins, dans ces conditions, on ne peut obtenir qu'un diagramme de diffraction partiel, sauf si l'on attaque l'échantillon sur l'un de ses bords. Toutefois l'analyse élémentaire étant donnée par la fluorescence, on peut interpréter convenablement ces diagrammes, qui représentent par ailleurs les meilleures conditions pour la mesure précise des paramètres.

Nous avons donc opéré en diffraction par la méthode photographique. Chaque diagramme correspond à l'irradiation d'une plage d'environ 0,5 mm.[2] En revanche, l'analyse élémentaire par fluorescence intéresse l'ensemble de la surface de la plus petite monnaie, la majeure partie de l'autre.

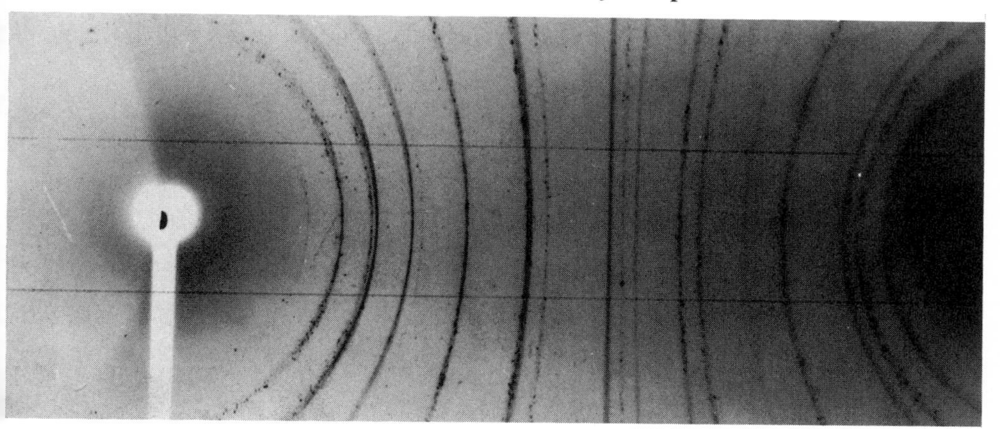

**Fig. 4. Diagramme de rayons X sur le bord de la pièce de bronze.**

Sur chaque pièce, un diagramme a d'abord été pris en un point quelconque de la circonférence, avec la radiation du cuivre excitée dans les conditions habituelles (35 à 40 kV, 5 mA, source fine), ce qui permet d'identifier les structures cristallines.

La comparaison des deux diagrammes ainsi obtenus

(figs. 4 et 5) illustre les avantages signalés au sujet de la méthode photographique. Le diagramme de la pièce de bronze (fig. 4) comporte deux séries de réflexions d'allures différentes: les unes sont ponctuées, les autres lisses et continues. Le dépouillement du cliché montre que celles de la première série correspondent à un oxyde, de structure type $Cu_2O$ (cuprite), celles de la seconde série indiquent une solution solide de paramètre voisin de celui du cuivre.

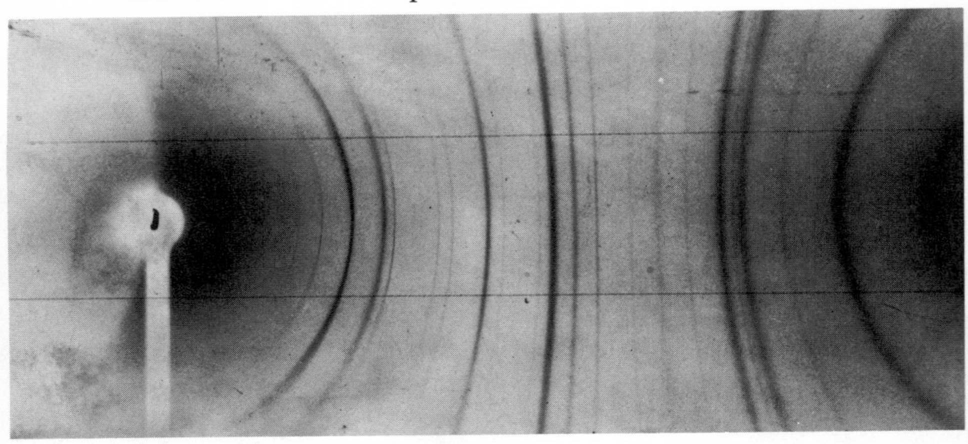

**Fig. 5. Diagramme de rayons X sur le bord de la pièce d'argent.**

Pour la monnaie gordienne (fig. 5), toutes les réflexions sont lisses. Les plus intenses, non résolues aux grands angles, correspondent à la structure de l'argent. L'ensemble des raies faibles, de même allure que les premières, signale l'oxyde d'argent, $Ag_2O$. Ici le diffractomètre à compteur aurait donné identiquement les mêmes informations. Au contraire pour le bronze, cuivre et oxyde donnent des raies à peu près de même intensité et la discrimination aurait été bien moins aisée si l'intégration par compteur avait fait disparaître les détails caractéristiques de chaque série. Notons au passage que ces métaux et leurs oxydes cristallisent tous dans le même système, mais avec des consmission. mètre que le tantes très différentes pour chacune des phases.

Les diagrammes en retour effectués tant sur la face que sur le revers de la pièce de bronze sont nettement différents suivant que la surface a été partiellement dégagée de l'oxyde qui l'encombrait (droit) ou examinée dans l'état de provenance (revers). Seules apparaissent les réflexions du cuivre lorsque la région irradiée est visiblement brillante (fig. 6 a), et on note leur dédoublement, caractéristique d'une bonne cristallisation. Pour effectuer avec exactitude la mesure du paramètre, il est commode de saupoudrer la surface de l'échantillon avec une substance étalon. Nous avons choisi le molybdène qui, en l'occurrence, permet une très bonne précision : l'une de ses réflexions coïncide en effet avec l'anneau ponctué de l'oxyde (fig. 6 b). On devine sur la reproduction, moins éloquente que le cliché original (fig. 6 b), les anneaux du cuivre en correspondance avec ceux de la figure 6 a, mais masqués par la gangue d'oxyde et l'absorption due au molybdène.

On trouve ainsi pour la maille de la phase cuivreuse un paramètre a = 3,619 Å au lieu de 3,6150 Å pour le cuivre pur. L'erreur expérimentale est inférieure à 0,003 Å du fait du calibrage. Néanmoins il est difficile d'affirmer que le cuivre contient ou non des éléments étrangers en solution. D'après l'analyse élémentaire, on est renvoyé au zinc —mais le taux serait faible, 2 à 3 %—ce qui ne veut pas dire qu'il n'y en ait pas eu davantage à l'origine. Nous y reviendrons : les transformations chimiques qui se sont produites au cours des siècles font partie de l'histoire de la pièce.

Pour l'oxyde de cuivre, la maille est de a = 4,266 Å au lieu de 4,2696 Å pour le minéral naturel et pur (cuprite). Il y a donc une légère contamination, que la seule mesure du paramètre ne permet pas de caractériser.

Cependant des renseignements complémentaires sur la nature de la corrosion vont être fournis par l'analyse des produits dégagés par les répliques. On obtient sur celles-ci d'excellents diagrammes par transmission.

Si l'on désire conserver intégralement la réplique, il faut toutefois se contenter des raies aux petits et aux grands angles, les réflexions moyennes étant absorbées par le support. Tel quel, le document suffit en général pour l'identification (fig. 7).

Sur la première réplique, on trouve des régions couleur de rouille, il s'agit là principalement, d'après le diagramme correspondant, de goethite impure (réf. 2–0281 de l'Index ASTM). Ailleurs, comme sur l'ensemble de la seconde réplique, on détecte un oxyde de zinc, $ZnO_2$, mêlé à des parcelles de cuivre et à du plomb accompagné de certains de ses oxydes.

Ces résultats amènent à proposer sinon une description de l'objet ancien, du moins un portrait reconstitué à partir de son dérivé recueilli—au vingtième siècle: une gangue aujourd'hui s'est formée où domine la rouille—ou goethite impure—témoignant d'une oxydation humide et sans doute d'un apport extérieur de fer. Sous cette première couche, ou mêlée à celle-ci, apparaît un produit de décomposition trés significatif: l'oxyde de zinc, avec du cuivre, du plomb et ses oxydes. Ce qui demeure de l'objet original est une masse cuivreuse, avec une teneur en zinc d'environ 3 pour cent au voisinage de la surface et une forte proportion d'oxyde de cuivre impur faisant corps avec la pièce.

Compte tenu des actions physiques et chimiques qui se

sont produites, des avatars que la pièce a subis, on peut donc proposer un classement métallurgique de cette monnaie : Il s'agit non pas d'un bronze classique, à base de cuivre-étain, mais d'un laiton au plomb. Ce laiton a perdu une partie du zinc, qui à l'origine était totalement en solution dans le cuivre, et se retrouve aujourd'hui partiellement à l'état d'oxyde. De plus, le plomb, comme les autres métaux, a formé ses propres oxydes. Les impuretés signalées par l'analyse par fluorescence (Tableau), sont vraisemblablement d'origine minérale : le plomb était chargé d'antimoine et d'argent, le cuivre apporta avec lui le nickel et le fer, peut-être aussi l'étain. Le taux original en zinc reste indéchiffrable à partir de l'examen superficiel d'une pièce corrodée.

Pour la pièce gordienne, non nettoyée et d'aspect gris verdâtre, il s'agit de savoir si la diffraction des rayons X permet la localisation des éléments signalés par la fluorescence. Nous supposerons au départ que le laboratoire ignore tout de ce type de monnaie et qu'il ne travaille qu'avec ses propres références.

La fluorescence indique une très forte proportion de cuivre (Tableau), cependant le diagramme pris sur le bord de la pièce paraît correspondre à celui de l'argent peu allié mélangé à l'oxyde.

Il semble donc à première vue que la pièce est saucée et qu'un mince revêtement d'argent couvre un noyau de cuivre-étain. Dans ce cas, et suivant les résultats de la fluorescence, la phase cuivreuse devrait apparaître relativement intense sur les diagrammes en retour, à côté de la phase à base d'argent.

Or ceux-ci pris en différents points des deux faces de la pièce ne présentent, outre les réflexions intenses de l'argent, que deux anneaux très faibles dont l'intensité relative varie d'ailleurs d'un cliché à l'autre. Du fait de sa mauvaise cristallisation, l'oxyde d'argent n'apparaîtrait pas aux grands angles dans les conditions où nous avons opéré, il s'agirait donc probablement des réflexions du noyau présumé, ceci avant tout dépouillement précis. En fait, la raie la plus pâle, fig. 8a, correspondrait à un plomb faiblement chargé en étain, le paramètre étant de 4,943 Å au lieu de 4,9506 Å pour le métal pur. L'autre réflexion serait celle d'une solution solide de cuivre, avec une maille

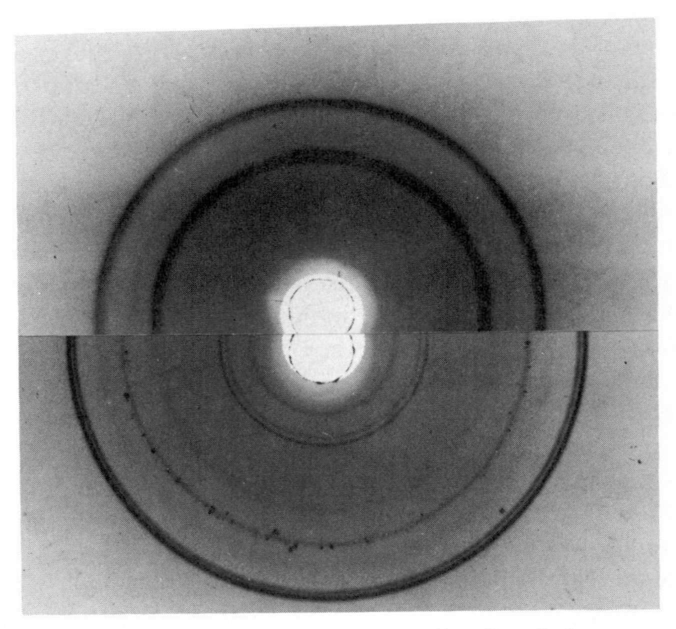

**Fig. 6. Diagrammes en retour sur la pièce de bronze.**
**a. Sur une parcelle de metal degagée d'oxyde.**
**b. Sur une région oxydee (avec en surimpression un calibrage à la poudre de molybdène).**

nettement dilatée: 3,638 Å (± 0,005 Å) au lieu de 3,6150 Å. L'élément dissous serait vraisemblablement l'étain; en négligeant les autres additions, le taux correspondant à ce paramètre serait de 8 pour cent.

Quant à l'argent, son paramètre est très voisin de celui du métal pur, soit 4,086 Å. Mais cette valeur n'implique pas qu'aucune trace de cuivre ou d'étain ne s'y trouve mêlée. Bien que nous ne possédions aucune donnée précise sur les alliages ternaires Ag-Cu-Sn, il est malgré tout permis de penser que certaines compositions one même paramètre que le métal pur, car, pris isolément, le cuivre contracte la maille tandis que l'étain la dilate.

S'il en était ainsi, ce qui nous est apparu comme "argent" pourrait être une phase ternaire, et les phases secondaires à base de cuivre ou de plomb ne seraient point sous-jacentes mais juxtaposées à cette solution solide à base d'argent.

Or, selon que l'on décide qu'il y a ou non revêtement d'argent, il s'agit de deux types de fabrication différente. L'analyse non-destructive permet-elle de prendre parti ?

Tentons de faire appel à l'aspect semi-quantitatif de l'analyse par fluorescence d'une part, de la diffraction X d'autre part.

En effet sur certains diagrammes en retour les réflexions du cuivre et du plomb sont à la limite de visibilité. Il se trouve donc là des épaisseurs d'argenture suffisantes pour absorber les réflexions de la radiation du cuivre sur le noyau si noyau il y a. Supposons que ceci corresponde à une réduction de l'intensité du faisceau de l'ordre de 600/1.[9] Dans ce cas, l'épaisseur x de la couche d'arrêt peut être évaluée par une formule simplifiée :

$$\mu x = 3{,}2$$

où $\mu$ est le coefficient massique d'absorption du revêtement.

Les tables ou ouvrages spécialisés [10] donnent les valeurs de $\mu/\rho$ (ou $\rho$ est le poids spécifique), le calcul—très approché—est donc extrêmement simple. On trouve que la couche limite d'argent qui arrêterait la réflexion du cuivre serait de 12,6 microns environ. Mais ici la radiation n'est totalement absorbée qu'en certaines régions, donc l'épaisseur d'argent ainsi calculée est une limite supérieure. Cependant le revêtement n'est pas de l'argent pur, il s'y mêle peut-être de l'étain dont l'absorption est du même ordre de grandeur que celle de l'argent, et éventuellement du cuivre, six fois moins absorbant que l'argent. Le rôle de l'oxygène est négligeable de ce point de vue. En définitive, si les réflexions attribuées aux solutions solides de cuivre ou de plomb étaient enregistrées après la traversée d'une couche d'argent, celle-ci serait tout au plus de 10 microns, car à partir de 12 ou 13 microns la visibilité des anneaux serait quasi nulle.

Comment concilier les résultats de ce calcul approché avec l'intensité des pics de fluorescence du cuivre? Le rayonnement primaire du molybdène provoquant cette émission comprend une importante fraction de haute énergie pénétrant bien au-delà de 10 ou 13 microns d'argent, mais une telle couche réduirait de 90 à 95 pour cent l'intensité de l'émission du cuivre à la sortie.

Il doit donc y avoir une importante proportion de cuivre située à une distance de la surface nettement inférieure à 10 microns. Cependant il a été également noté qu'aucune impureté de la mine du cuivre n'est signalée par l'analyse en fluorescence—pas plus d'ailleurs que celles qui généralement accompagnent le plomb et l'antimoine. Ici intervient un effet de quantité. S'il y avait relativement peu de cuivre au voisinage immédiat de la surface, les faibles émissions correspondant aux traces de nickel ou de fer par exemple seraient absorbées par l'argent environnant.

et la nature des résultats escomptés. Mais ceci, nous avons essayé de le montrer, dépend en fait du type d'échantillon, de son volume et de son état de conservation d'une part, de l'état d'avancement de l'inventaire de notre savoir d'autre part.

La pièce de monnaie représente le cas particulier d'un objet qui peut être soumis à des observations au microscope, sous réserve toutefois d'une préparation de surface et des inconvénients dus au relief. De ce point de vue elle se prête moins bien à l'examen microscopique superficiel qu'un bronze de grand volume, comportant des régions non destinées à être contemplées. Sur celles-ci on peut au moins pratiquer un examen métallographique non destructif [15] qui évite le rapprochement de l'objectif du microscope et de l'objet d'art, donc son transport au laboratoire. De même, aussi petite que soit l'étincelle du spectrographe dans l'ultra-violet elle est plus préjudiciable à une pièce de monnaie, couramment étudiée à la loupe, qu'à une pièce d'orfèvrerie comportant des parties cachées.

Ainsi, suivant un premier classement, l'objet sera passible ou non de méthodes d'observation ou d'analyses qui ne sont pas rigoureusement non-destructives, selon notre définition première.

Considérons maintenant un tout autre aspect de la question, celui de l'état de conservation de l'objet. Du point de vue métallurgique, la probabilité qu'un objet de fabrication ancienne nous soit livré dans des conditions représentatives de son état d'origine est relativement faible. Ce qui fausse notre perspective lorsque le phénomène n'est pas aussi apparent que celui d'une oxydation ou d'une décomposition de l'alliage pouvant aller jusqu'à la décohésion totale, c'est l'état exceptionnel de conservation de certains témoins. Cependant des ouvrages comme celui de Cyril S. Smith [16] montrent bien qu'à chaque moment de l'histoire des sciences, les progrès de l'observation ou de l'analyse, appliqués à des objets d'un passé lointain ont fait soupconner l'intervention d'un facteur jusqu'alors inconnu.

Ainsi nous acceptons beaucoup mieux aujourd'hui qu'il y a seulement vingt-cinq ans l'idée d'une diffusion intermétallique dans une pièce gordienne. Non que nous ayons pénétré profondément le mécanisme de cette diffusion, mais parce que nous savons qu'elle existe, et que nous disposons de méthodes, comme celle des traceurs radioactifs, pour l'étudier lorsqu'elle s'est produite. Si quelque jour les lois de la diffusion étaient découvertes, non seulement des progrès considérables pourraient être réalisés dans les constructions les plus ambitieuses de notre époque, mais nous pourrions alors reconstituer beaucoup plus exactement les conditions dans lesquelles les alliages de l'antiquité ont été élaborés. Car les phénomènes fondamentaux n'ont pas d'âge, et les techniques les plus primitives sont soumises aux mêmes impératifs que les alliages à notre sens les plus raffinés.

Ce qu'on appelait à tort les secrets du passé dans la littérature du dix neuviéme siècle, c'est le début de la conaissance d'un phénomène dont l'objet antique apporte la révélation, sans que ses auteurs n'en aient sans doute été avertis. Encore aujourd'hui, ce qui nous fait le plus cruellement défaut pour reconstituer le passé d'un objet, pour retrouver les conditions de son élaboration première, ce sont les lois d'évolution à long terme des matériaux métalliques. Si nous sommes relativement bien avertis des phénomènes de corrosion chimique qui se manifestent à l'échelle macroscopique, il nous reste à découvrir les lois de la diffusion à long terme, facteur sournois d'hétérogénéisation des alliages les plus divers. Faute de données théoriques suffisantes à ce sujet, nous en sommes réduits à livrer contre le temps un combat sans issue. Même si nous voulions préparer la tâche de nos lointains successeurs en leur livrant des cahiers de laboratoire impeccables à nos yeux, nous serions bien présomptueux en espérant satisfaire leurs exigences de l'an 4.000.

Avant même d'aborder le problème de l'historien, celui qui concerne l'âge réel de l'objet par rapport à l'acquisition des connaissances scientifiques ou techniques dont il témoigne, le laboratoire doit retrouver l'âge intrinsèque du métal qu'il manipule.

Toute analyse banale suppose connus les éléments préalables à l'entrée au laboratoire du spécimen voué à l'analyse, mais bien des éléments manquent encore pour reconstituer à coup sûr le passé d'un alliage, s'il ne s'agit ni d'une fabrication courante, ni d'une existence bien plus brève que celle de l'expérimentateur; et de ceci nous sommes chaque jour plus avertis.

L'examen de l'objet ancien exige donc à la fois un effort de réflexion inhabituel, et des connaissances scientifiques plus étendues, dépassant parfois largement celles nécessaires aux opérations de routine. Chaque analyse banale est fondée sur un bagage représenté par des tables numériques confirmant des lois générales ou en tenant lieu. En physique des métaux, on ne dispose que d'un nombre restreint de lois générales. Si les associations binaires courantes sont bien connues, d'autres, plus exceptionnelles sont partiellement représentées ou manquent totalement dans les catalogues, et ceci devient le cas général pour les alliages ternaires ou plus complexes. De plus le rôle des oligo-éléments se révèle plus important au fur et à mesure que nous savons mieux en chiffrer les traces, et ceci peut jouer même dans l'évolution à court terme de nos fabrications les plus soignées.

Il y a donc de sérieux problèmes intérieurs au laboratoire au sujet de l'interprétation même de l'analyse et on ne peut guère invoquer de doctrine générale pour les résoudre, on ne peut manquer de constater au contraire que chaque nouvel effort fait entrevoir les lacunes à combler.

Ce qui paraît raisonnable, c'est donc de procéder en quelque sorte par approximations successives, et en tout cas de commencer pour l'objet de nature inconnue par les analyses non-destructives, au moins pour dégrossir les questions, classer sommairement l'objet du point de vue métallurgique, en distinguant les résultats "normaux" de ceux qui sont "insolites," quitte à laisser subsister quelques points d'interrogation.

Nous avons vu, par exemple, que le "moyen-bronze" romain est à vrai dire pour le métallurgiste, un laiton au plomb. La méthode de l'analyse sur empreinte a montré que la dézincification s'accompagnait de la formation d'un oxyde ne faisant pas corps avec la masse de l'objet, où au contraire oxyde de cuivre et cuivre semblent intimement mêlés, quoique de manière hétérogène.

Veut-on aller plus loin, il faut alors recourir à d'autres méthodes, poser la question de la destruction de l'objet, suivant sa valeur propre, sa rareté, et la nature des problèmes que les premières analyses permettent de poser avec précision. Pour la pièce gordienne, par exemple, l'hétérogénéité en soi a été révélée par des examens superficiels non-destructifs (diffraction X), mais la manière dont elle s'est produite ne peut être étudiée que par des examens microscopiques étayés par des mesures à la microsonde électronique. C'est là le type d'un problème ardu que seuls les progrès les plus récents de la micrographie donnent aujourd'hui l'espoir de pouvoir aborder.

Mais il faut réserver raisonnablement à des cas de ce genre, que l'analyse rapide révèle avant destruction, les efforts qu'implique une recherche systématique et délicate. En général l'analyse rapide permettra donc de classer les problèmes, et ce n'est pas là le moindre de ses mérites : elle dirige en quelque sorte la propre économie de l'analyste.

En ce qui concerne les traces d'éléments insolites, l'analyse qualitative est extrêmement fructueuse en bien des circonstances, et il convient de la pousser aussi loin que possible. Ainsi, nous avons signalé que la bague d'or de la reine mérovingienne Arnegonde contenait outre 4 pour cent d'argent environ, et approximativement 2 pour cent de cuivre, une moindre proportion de cobalt, sans parler des impuretés de la mine. Encore s'agit-il de savoir si cette impureté de surface n'est pas une contamination du creuset. Du moins y-a-t-il là une indication, comme ailleurs la présence du cadmium ou du strontium dans l'argent ou le plomb.

L'analyse quantitative a été utilisée dans le même sens, nous l'avons vu, par Caley mais il s'agit de bien réfléchir à son utilité avant d'y recourir, car elle ne se justifie pleinement que lorsqu'une méthode de filiation est significative à coup sûr. On peut citer les cas où l'historien recherche une dévaluation monétaire, ceux où l'archéologue affirme qu'il n'a pas affaire à une tribu nomade ayant éparpillé ses oeuvres dans un territoire plus ou moins vaste[17]; ceux enfin où il s'agit de préciser la signification technique d'un mode particulier d'hétérogénéité.

Ce ne sont là que des exemples, destinés à illustrer le mode opératoire que commande le bon sens le plus élémentaire. Il paraît s'en dégager deux règles principales : dégrossir le problème sans détruire, ne pas chercher à doser ou à raffiner un dosage sans savoir exactement pourquoi.

De nos jours l'objet fabriqué par l'ancêtre plus ou moins lointain de nos industriels et artisans contemporains n'est toutefois pas le seul à pénétrer au laboratoire sans acte de naissance, ce qui *a priori* lui confère un titre de non reproductibilité sinon de rareté. Plus jeune que l'échantillon géologique, l'objet ancien en métal obéira néanmoins à cette loi inéluctable énoncée par U. R. Evans au début de son magistral ouvrage sur la corrosion : tiré du minerai par l'homme, le métal livré à lui-même tend à se minéraliser. Le vieillissement même de l'objet se ramène donc à l'association d'une géologie abrégée avec une métallurgie étendue.

Mais le laboratoire ne borne plus aujourd'hui son ambition à l'analyse des produits extraits de la croûte terrestre ou à ceux des premiers forgerons. Les espaces interplanétaires livrent à leur tour des débris d'astres qui sont autant d'échantillons pour l'analyste. Ces fragments de météorites, où une fois encore l'inventaire analytique comporte des éléments classiques et des trouvailles insolites, posent avec des dimensions différentes—au temps s'ajoute l'espace—des problèmes sinon analogues, du moins de même nature que ceux envisagés au sujet des métaux anciens.

Tout manque ici, une fois encore, sauf l'épilogue, et l'étude même de la contamination n'est pas la moindre partie de l'histoire de ces météorites, dont il faut reconstituer le voyage interplanétaire, avant et après la traversée de l'atmosphère terrestre et le choc avec notre globe. Mais ce qui intéresse directement notre propos, toutes proportions gardées, ce sont les règles qui s'imposent au sujet de leur étude au laboratoire, en ce qui concerne d'une part ce qui demeure incônnu quant à l'origine, et aux péripéties suivantes, d'autre part relativement à ce qui peut être raisonnablement escompté des analyses. Ici plus encore que pour la plupart des objets, les humains (les terriens) ne disposent d'échantillons qu'en quantités relativement limitées : à nouveau un système d'économie s'impose à l'analyste.[18] Le problème alors dépasse et notre planète et la chronologie terrestre. Il s'agit de savoir si les résidus carbonés apportent la preuve d'une vie organique extra-terrestre, et si ces embryons ont un âge.

Ainsi l'analyse est aujourd'hui une discipline de grande ambition. Du même coup, elle cesse d'être simple dissection pour devenir source de réflexion et sur le savoir même qu'elle applique, et sur les connaissances qui se sont progressivement accumulées et dont chaque échantillon porte le message.

TABLEAU

Résultat des Analyses Qualitatives par Fluorescence X

| | *Pièce de bronze* | *Pièce argentée* | *Proportions estimées* |
|---|---|---|---|
| *Eléments de base* | Cuivre | Argent<br>Cuivre | <br>Ag/Cu = 60/40 |
| *Additions intention-nelles* | Zinc<br>Plomb | Etain<br>Plomb | Ag/Sn = 100/20<br>Sn/Pb = 20/10 |
| *Impuretés* | Antimoine<br>Fer<br>Nickel | | |
| *Traces* | Etain<br>Argent | Or (non exclu)<br>Fer (absent)<br>Antimoine (absent)<br>Zinc (absent) | |

## REFERENCES

1. A. R. Weill, Communication à la réunion de l'Institut International pour la Conservation des objets d'art et d'histoire (ROME, 1961); *Recent Advances in Conservation,* Butterworth, 1962. p. 111.
2. A. R. Weill, *Conservation* (Ehides de) (1), 30, 1952.
3. A. R. Weill, *Mém. Sci. de la Revue de Métallurgie*—57 (6), 459, 1960.
4. L. Gramme et A. R. Weill, ibid. (7), 524, 1952 et 51 (7), 459, 1954.
5. N. F. M. Henry, H. Lipson et W. A. Wooster *The Interpretation of X-Ray Diffraction Photographs* (Londres: MacMillan & Co., 1951).
6. Y. Cauchois et H. Hulubei, *Longueurs d'onde des émissions X et des discontinuités d'absorption X* (Paris: Hermann & Co., 1947).
7. F. Gaillard et A. R. Weill, *Mém. Sci. de la Revue de Métallurgie* 57 (12), 889, 1960.
8. A. R. Weill, *Bulletin du Laboratoire du Musée du Louvre.* No 4, Sept. 1959, p. 21.
9. A. Taylor, *An Introduction to X-ray Metallography* (Londres: Chapman & Hall Ltd., 1949), p. 364.
10. A. Taylor, *X-Ray Metallography* (New York: John Wiley & Sons, 1961).
11. E. T. Hall, Communication à la réunion de l'Institut International pour la Conservation des objets d'art et d'histoire (Rome, 1961); *loc. cit.* (1) p. 29.
12. J. Guey et J. Condamin. *Revue Numismatique,* VIe série, III (1961), p. 51.
13. E. R. Caley—*Ohio J. Sci.* 61 (1961), p. 151; d'après *I. I. C. Abstract* III, (4) n° 3293.
14. W. W. Meinke et R. W. Shideler—*Nucleonics,* 20 (3), 1962, p. 60.
15. P. A. Jacquet, "Technique non-destructive pour les observations, en particulier de nature métallographique, sur les surfaces métalliques," Note Technique N° 54 (1959), Office National d'Etudes et de Recherches Aéronautiques (Châtillon-sous-Bagneux, Seine, France).
16. C. S. Smith, *A History of Metallography* (Chicago: University of Chicago Press, 1961).
17. F. Tavadze et Th. Sakvarelidze, *Bronzes of Ancient Georgia* (en géorgien, résumé en anglais) Tiflis, 1959.
18. J. D. Bernal, *Nature,* 190 (1961) p. 129.

# Pigments Employed in Old Paintings of Japan

By Kazuo Yamasaki
Nagoya University
Nagoya, Japan

## INTRODUCTION

The first chemical study on pigments employed in old paintings of Japan was probably that on the seventh century wall painting of the Hôryûji temple near Nara carried out in 1919 by the late Professor Chikashige. After a lapse of two decades the study on this wall painting was resumed again in 1939 by Professor Y. Shibata and the present author, and the research has been extended since then to include paintings of the protohistoric time and also those of the periods later than the Hôryûji painting. In this article, a survey on the pigments used in Japanese painting will be presented.[1] The method of investigation included microchemical and spectrochemical analyses, ultraviolet and infrared photography and X-ray radiography. Application of $\beta$-ray back-scattering to the study of pigments was also tested.

## PRE-BUDDHIST PAINTING

The early cultural periods in Japan are divided into two, the Jômôn and the Yayoi periods, both named after types of excavated pottery. The former, the period of hunting and fishing, lasted until about the third century B.C. Its date of beginning is not certain yet, several thousand years of duration being claimed by the radiocarbon test. The Yayoi period, the period of agriculture and metal-working, lasted from the third century B.C. to about third century A.D. In these two periods, however, there was no pictorial art in color, except for some pottery painted with red ochre or cinnabar, and reliefs on bronze bells.

After the Yayoi period there came an era when huge mounds were constructed for the burial of the dead. Thus it is called the Kofun period, the period of tumuli. This period lasted from about the third century to the end of the seventh century, and in its latter part overlapped with the historic Asuka-Nara period without clear demarcation.

Some of the tombs of the fifth and sixth centuries in the Kofun period were decorated with paintings; these tombs, having decorated funerary chambers, are confined almost exclusively to the northern part of Kyûshû, i.e. the Fukuoka and Kumamoto prefectures.

The present author [2] studied about forty of these tombs including all the important ones on which archeological studies were made. Most of the tombs have only simple decorations such as concentric circles, or triangles painted in red alone, but some of them have figures like armor, men, horses, birds, boats, in addition to the geometrical designs. The most gorgeously decorated tomb is the Otsuka tomb situated in Keisen-machi, Fukuoka prefecture.[3] The

a.

b.

**Fig. 1. The wall painting of the Takehara tomb (a) and its sketch (b). The latter was reproduced from an article by T. Mori.**

paintings cover the entire walls of the funerary chamber, representing horses, men, quivers and swords, triangles and mysterious geometrical figures. Five colors are used: white, yellow, red, green and black (cf. Table 1). Another tomb which has quite an interesting painting is the Takehara tomb situated at Wakamiya-machi, Fukuoka prefecture.[4] Its date is the late sixth century. The painting consists of only two colors; red ochre and carbon. As seen in Fig. 1, a man holding a horse stands in the center and a queer animal, a dragon-horse, is above. Two boats, two sunshades and an ensign are shown with wave-like forms at the bottom. This picture may be interpreted as representing the travel of the dead man's soul to the next world by a horse on land, by boats on sea and by a dragon-horse through the sky. The kinds of pigments found so far in these decorated tombs are listed in Table 1.

TABLE 1.

| *Color* | *Pigment* |
|---|---|
| white | clay |
| red | impure red ochre |
| yellow | impure yellow ochre |
| green<br>bluish green | powdered green rock |
| black | carbon<br>black mineral containing manganese and iron |

There are two kinds of black pigment: carbon and a black mineral containing manganese and iron. Table 1 also shows the interesting fact that no pigments like azurite, malachite, and cinnabar were used and only colored materials found near the tombs were employed. Another fact is the problem of the ground coating. In tombs of the Koguryo period in Korea and those of the Han period found in the northeastern part of China, lime was used as the ground coating and it has been changed to calcium carbonate during thousands of years, but no such case has been found in the decorated tombs in Kyûshû, in which pigments were applied directly on stone surfaces.

## BUDDHIST PAINTING

*(1) Paintings of the seventh-eighth centuries.* In 538 A.D. a Korean emperor presented some sutras and a Buddhist statue to Japan, and this was the official introduction of Buddhism. It spread gradually into the country and many Buddhist temples were built. Among them the Hôryûji temple near Nara which is the oldest one surviving to the present day and preserves many examples of early Buddhist art such as the well-known wall painting. In addition to the Hôryûji painting about seventy wall paintings made in various periods are now extant in Japan, mostly in Buddhist temples. All of them are painted on wooden panels or doors except for three: a mud wall-painting of the Hôryûji and two others on walls of lime plaster.

The inner walls of the sanctuary of the Golden Hall of the Hôryûji were decorated with brilliant paintings of the late seventh century which were severely damaged by a fire in 1949. These paintings consist of four large units (ca. 3m x 2.6m) and eight smaller ones (*ca.*3m x 1.5 m). Each of the large walls represents a Buddhist paradise and the small wall a figure of Bodhisattva. The wall itself is made of a mixture of mud, sand and straw bits as a binder. The surface of the walls is coated with fine white clay upon which are painted the Buddhist figures with the mineral pigments listed in Table 2.[5] Though adhesive animal glue must have been used, and there is documentary evidence that it was, it is difficult to prove its use now.

TABLE 2.

| *Color* | *Hôryûji late 7th century* | *Daigoji pagoda 951* | *Hôwôdò (Phoenix Hall) 1053* | *Sliding doors 17th century* |
| --- | --- | --- | --- | --- |
| *White* | clay* | clay* white lead | clay* | calcium carbonate |
| *Red* | cinnabar red ochre red lead | same | same | same |
| *Yellow* | litharge yellow ochre | litharge gold | litharge yellow ochre gold | yellow ochre gold |
| *Green* | malachite | malachite | malachite | malachite |
| *Blue* | azurite | substitute for azurite | azurite substitute for azurite | azurite |
| *Black* | Chinese ink | same | same | same |

* Clay is used as the ground coating, too.

It should be noted that these mineral pigments, still in use today, are quite different from those employed in protohistoric tomb paintings. The artificially made pigments like red lead, litharge and bright-colored minerals like azurite and malachite have come into use. Red ochre, too, was purer in the Hôryûji painting than in the tomb paintings. The introduction of Buddhism brought with it a new technique and new materials on which Japanese painting has been based ever since. Some documentary evidence also support this view. In 610 a monk from Koguryo named Donchô introduced to Japan the technique of preparing pigments and painting materials. In 692, another monk Kanshô was rewarded with a prize for preparing white lead.

Quantitative chemical analyses were not possible of the pigments used in the Hôryûji painting. The red ochre samples, however, applied on the wooden parts of the pagoda of the Hôryûji which were built at nearly the same time as the Golden Hall were available for analysis. The data are given in the Table 3.[6] Under the microscope the samples were seen to contain quartz and limonite. The spectral reflectance curves were measured and the dominant wavelength, purity and luminosity calculated on the CIE system are listed in Table 3.

Besides the pigments actually used in the Hôryûji painting a few pigments of the eighth century have remained in the Shôsôin. The Shôsôin, the famous Repository of Imperial Treasures in Nara was originally one of the storehouses belonging to the Tôdaiji temple, the temple of the Great Buddha. In 756, four years after the completion of the Great Buddha, the Emperor Shômu, who had made it, died, and his widow, the Empress Kômyô, donated more than six hundred precious objects to the Great Buddha as an offering. These constitute the nucleus of the Shôsôin

Treasures which have been preserved with great care to the present day. The fact that a large number of relics 1,200 years old have been preserved nearly intact together with original labels and lists is almost unparalleled in the world's history. Among the objects dedicated to the Great Buddha by the Empress Kômyô were sixty items of medicine. Among these medicines were red lead and silver powder. But although preserved together with medicines such as rhubarb, ginseng, cinnamon and Epsom salt, it appears that these were not used as medicines, but as painting materials.

TABLE 3.

| | *Red ochre No. 3, 7th century* | *Red ochre No. 6 date uncertain, but older than 1280* |
|---|---|---|
| *Loss on ignition* | 17.22% | 12.42% |
| $SiO_2$ | 30.26 | 42.44 |
| $Fe_2O_3$ | 41.71 | 30.50 |
| $Al_2O_3$ | not determined | 11.66 |
| *MgO* | n. d. | 3.24 |
| *Total* | 89.19% | 100.26% |
| $\lambda d$ | 597 m$\mu$ | 597 m$\mu$ |
| *Y* | 7.7% | 6.0% |
| *Pe* | 38.2% | 47.5% |
| *Munsell re-notation* | 9.8R 3.3/4.2 | 1.0YR 2.9/4.6 |

The red lead [7] is in powder form and amounts to about 100 kilograms, contained in 128 small paper packages. On the undersides of the outer wrappings of the packages, the weights and the grades "superior," "medium," or "low," written at the time of original packaging. These red lead samples consist of fine orange-colored particles with diameters of 0.005–0.01 millimeters. The chemical composition and the color calculated on the CIE system are given in Table 4. As clearly shown in the table, the samples are very low in $Pb_3O_4$ as compared with modern red lead.

Silver powder [8] has the diameter of 0.05–0.5 millimeters with the chemical composition Ag 94.17, Au 1.14, Cu 0.84, substances insoluble in aqua regia, 0.57 percent, total 96.72 percent. Figures less than 100 percent mean oxygen in the sample and undetermined components. Besides the above elements, the following were detected by spectrographic analysis: Si, Mg, Fe, Mn, Al, Pb and Ca.

TABLE 4.

| | *Superior* | *Medium* | *Low* | *Modern red lead* |
|---|---|---|---|---|
| $H_2O$ | 0.06 | 0.06 | 0.08 | — |
| Acid-insoluble | 0.35 | 0.42 | 0.24 | 0.48 |
| Pb metal | 0.82 | 0.63 | 0.58 | — |
| $Pb_3O_4$ | 26.20 | 23.98 | 6.17 | 78.10 |
| PbO | 70.90 | 73.00 | 92.20 | 21.10 |
| Ag metal | trace | trace | trace | — |
| $Fe_2O_3$ | 0.01 | 0.02 | 0.02 | — |
| $Al_2O_3$ | 0.23 | 0.54 | 0.37 | — |
| CuO | 0.03 | 0.01 | 0.04 | — |
| CaO | 0.18 | 0.12 | 0.12 | — |
| MgO | 0.03 | 0.04 | 0.01 | — |
| Total | 98.81% | 98.82% | 99.83% | 99.68% |
| $\lambda d$ | 595 | 595 | 593 | 602 m$\mu$ |
| Y | 20.8 | 23.0 | 23.1 | 20.4% |
| Pe | 62.5 | 69.5 | 51.8 | 74.2% |
| Munsell re-notation | 1YR 5.1/9.2 | 1YR 5.3/10.6 | 1.3YR 5.4/7.4 | 9R 5.1/13.2 |

*(2) Wall paintings of the later periods.* In addition to the wall painting of the Hôryûji temple, pigments of about forty other wall paintings of the later periods were studied. Some of the results are listed in Table 2. Changes in kinds of pigments employed throughout the long periods ranging from the seventh to the nineteenth century can be said to

a.

Fig. 2. A part of the board covering the center pillar of the pagoda of the Daigoji temple (a) and its X-ray radiograph (b). The bracelets are painted in white lead impenetrable to X-rays and the pigment at the knees has fallen away. (Courtesy of the Tokyo Bunkazai Kenkyusho.)

b.

be rather few. The most significant change is in the use of powdered shell, calcium carbonate, as the white pigment instead of clay. This change seems to have taken place at some time in the fifteenth to the sixteenth century. The accurate determination of the date of change is difficult, as the examples of wall paintings with exact date and without retouching are very few during that period.

Another change is the use of a substitute for azurite in the tenth to eleventh centuries. This substitute for azurite made by dyeing yellow ochre with indigo was first found in the wall painting of Hôwôdô (Phoenix-Hall, 1053 A.D.) at Uji near Kyôto.[9] The original blue color is almost lost now, but from the margin of the painting kept concealed under the frame a bluish color appeared during the repair work and indigo was identified. The same substitute for azurite has been found in the painting of the pagoda of the Daigo-ji temple (951 A.D.)[10] and also in the wooden halos of several statues in the eleventh century, but this substitute has not been mentioned in the documents concerning pigments. The reason why this substitute was used is not known. Perhaps the production of azurite or its import from abroad was difficult in the tenth to eleventh centuries. In the twelfth to thirteenth centuries, azurite was again used in wall paintings and after the sixteenth century the pigments used were nearly the same as in modern Japanese painting, *i.e.,* cinnabar, red lead and red ochre for the red; malachite for the green; azurite for the blue; yellow ochre for the yellow; calcium carbonate for the white; gold and silver.

## SCREEN AND DOOR PAINTINGS

In the sixteenth century when the warriors like Oda Nobunaga and Toyotomi Hideyoshi won and gained political power, a grandiose style of painting was born. Many pictures were painted on sliding doors and folding screens of newly built castles and palaces. The themes were landscape, birds and flowers, not Buddhist motif. In these paintings on paper, gold was profusely used in powder and leaf forms, and pigments were applied in thick layers. The palette of the painters of these screen and door paintings were nearly the same as those of modern painters.

## SCROLL PAINTING

In addition to the wall, door and screen paintings, there is another form of painting which should be mentioned—scroll painting. One of the early examples is the E-ingakyô (illustrated sutra of Causes and Effects) of the early eighth century.[11] Another example described here is the Genji Monogatari Emaki (Scroll of the Tale of Genji) of the early half of the twelfth century.[12]

In examining pigments used on scroll paintings made of paper, X-ray radiography proved to be most useful, because these paintings are too precious to be submitted to chemical examination. On examination of the above two scrolls, some pigments not used in other kinds of paintings are found. These are gamboge, and an unidentified red organic pigment which is supposed to be of plant origin. The results are listed in the Table 5.

Another method of identifying pigments without harming the painting is by the use of back-scattering of beta particles emitted from radio-isotopes.[13] When beta particles strike an object, most of them are absorbed, but some particles rebound. This is the phenomenon of back-scattering, and the intensity of back-scattering increases with the atomic number of the elements constituting the object.

TABLE 5.

| Color | E-ingakyô | Tale of Genji |
|---|---|---|
| *White* | white lead | white lead<br>clay (uncertain) |
| *Red* | cinnabar<br>red lead | cinnabar<br>organic pigment |
| *Green* | malachite | malachite |
| *Blue* | azurite | azurite |
| *Black* | Chinese ink | Chinese ink |
| *Others* | — | gold<br>silver |

Back-scattering may be applied to the identification of pigments. A pigment such as white lead gives a high intensity of back-scattering of beta-particles, while calcium carbonate and clay give smaller values. In the same way. cinnabar may be easily distinguished from red ochre. These are the results of preliminary experiments; finer technical improvements are necessary for determination of substances which are made of elements having atomic weights of the same order.

## OIL PAINTING

In the treasures preserved in the Shôsôin, there are various kinds of handicrafts and paintings in which oil is used as the medium for pigments. The use of oil painting on art objects of the seventh and eighth centuries is most surprising from the standpoint of the history of oil painting in Europe, where the origin of oil painting is supposed to date from about the twelfth century.

There are two kinds of techniques in the use of oil among the treasures of Shôsôin.[14] One is the true oil-painting technique in which the mixture of pigments with oil is used, and the other is the application of oil over paint-

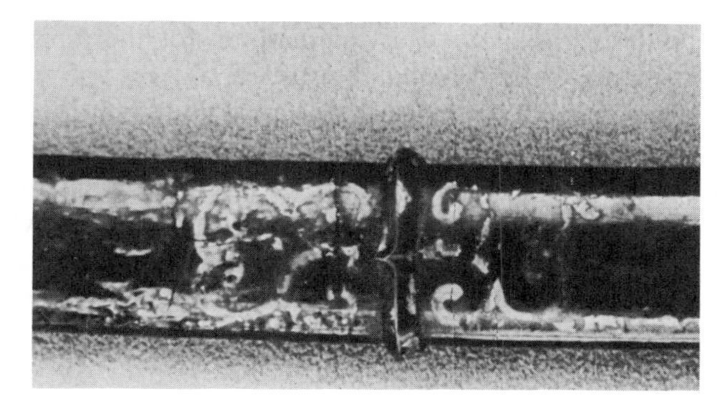

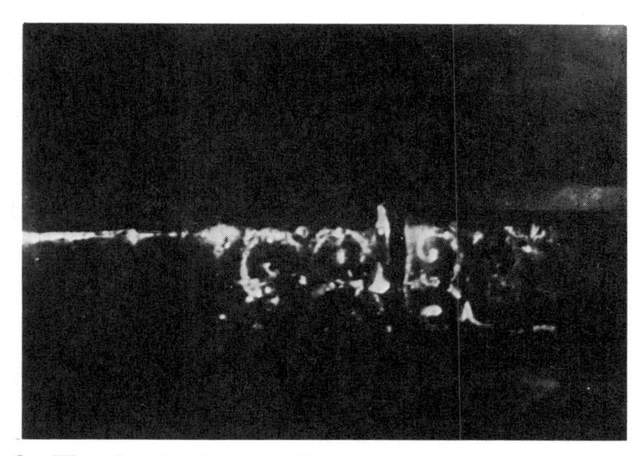

**Fig. 3. The sheath of a sword preserved in the Shosoin (above) and its fluorescence under ultraviolet light (below). (Courtesy of the Shosoin office.)**

ing, in which glue is used as the medium for the pigments. The evidence which proved the use of oil was the yellow fluorescence which appeared under ultraviolet light. For example, on the sheath of a sword covered with Japanese lacquer, patterns are painted with a white pigment mixed with oil. Under ultraviolet light, only those parts where oil was used show fluorescence since Japanese lacquer does not emit fluorescence. The oil used for these paintings may be supposed to be perilla oil, judging from the documentary evidence.

One of the reasons why oil was used may have been the difficulty of obtaining colored Japanese lacquers by mixing with pigments. Almost all the pigments except cinnabar are blackened when mixed with Japanese lacquer, while original colors of the pigments are retained when mixed with oil. Pigments used in the oil paintings of Shôsôin are clay, white lead, cinnabar, red ochre, red lead, yellow ochre, malachite, azurite, gold, and silver so far studied.

These oil paintings are now called Mitsuda-e, because perilla oil is boiled with *Mitsudasô* (Japanese word from the Persian *Murdasang*), litharge, and it gave the oil a drying property. So litharge was used as a siccative, but not as the pigment to be mixed with oil. The use of lead compounds as the siccative was proved by the presence of lead in the oil film, taken from the scroll roller of a sutra of the eighth century, which was revealed spectrographically. Besides the Shôsôin treasures, a Buddhist shrine, Tamamushi-zushi, preserved in the Hôryûji also has oil paintings which, dating from the mid-7th century, is probably the oldest example of the Mitsuda-e in Japan.

The history of oil painting in the East seems to be traceable back to the pre-Han period in China by documentary evidence, although no relics of that period have so far been found.

## COMPARISON OF JAPANESE WALL PAINTINGS WITH OTHERS IN THE EAST

Comparative studies of pigments used in the ancient paintings in Japan with those in other countries such as China, Central Asia and India seem to be interesting, but only a few results have so far been published. Nevertheless a glance at Table 6 reveals an interesting fact about blue and green pigments.[15] In Bamiyan and Qyzyl paintings, lapis lazuli (natural ultramarine) and chrysocolla were used for the blue and the green pigments respectively, while in Bezeklik and Tun-Huang azurite and malachite were used. The use of lapis lazuli seems to be closely related to the geographical location of the wall paintings and it may be that the Bezeklik painting was under the same cultural influences as the Tun-Huang painting. According to art historians, the Bezeklik painting is of the Chinese Buddhistic style. This view is in accord with the pigment studies.

By comparing Tables 2 and 6 it may be said that characteristic points of the Japanese wall painting are the use of clay as the ground coating and of azurite as the blue pigment.

This article is dedicated to Professor Yuji Shibata in commemoration of his eightieth birthday. Under his guidance the present author began the study of archeological chemistry 25 years ago.

Inorganic Chemistry Laboratory
Nagoya University
Nagoya, Japan

TABLE 6.

| *Wall Painting* | *Ground Coating* | *Red* | *Yellow* | *Blue* | *Green* | *White* | *Black* | *Reference* |
|---|---|---|---|---|---|---|---|---|
| 1. *Ajanta* (*5-6th c.*) | gypsum lime | red ochre | yellow ochre | lapis lazuli | terre verte | — | carbon | 16 |
| 2. *Bamiyan* (*5-6th c.*) | gypsum over mud | red lead red ochre | yellow ochre | lapis lazuli | chryso-colla | gypsum | carbon | 17 |
| 3. *Qyzyl* (*7th c.*) | gypsum | red lead red ochre | — | lapis lazuli | chryso-colla | gypsum | — | 18 |
| 4. *Bezeklik* (*9-10-11th c.*) | gypsum lime | red lead red ochre cinnabar | yellow ochre | azurite | malachite | — | carbon | 19 |
| 5. *Miran* (*3-4th c.*) | gypsum | cinnabar | yellow ochre | — | malachite | gypsum | carbon | 20 |
| 6. *Tun-Huang* (*8-9th c.*) | clay | red lead red ochre cinnabar organic pigment | gamboge | azurite indigo | malachite | clay white lead | Chinese ink | 21 |
| 7. *Wan Fo-Hsia* (*16th c.*) | clay | red lead red ochre cinnabar | yellow ochre | azurite | malachite | clay | Chinese ink | 21 |
| 8. *Ching-Ling, Mongolia* (*11th c.*) | lime | red lead red ochre cinnabar | yellow ochre gold | azurite | malachite | — | Chinese ink | 22 |

## NOTES

1. For the general survey of Japanese painting, see T. Akiyama, "Japanese Painting," (Switzerland: Skira, 1961).
2. K. Yamasaki: "Sci. Papers of Japanese Antiques and Art Craft," * No. 2, 8(1951), and unpublished results.
3. S. Umehara and Y. Kobayashi, "Report on Archeological Research in the Dept. of Literature,' Kyôto Univ., vol. 15(1939).
4. T. Mori, *Bijutsu Kenkyû* ("J. Art Studies"),* No. 194, 1(1957).
5. Y. Shibata and K. Yamasaki, *Proc. Japan Acad.,* 24, No. 2. 11(1948), K. Yamasaki: Bijutsu Kenkyû,* No. 167, 32(1952).
6. K. Yamasaki and N. Ohashi, *Sci. Papers Japanese Antiques and Art Crafts,* No. 5, 7(1953).
7. K. Yamasaki, *Studies in Conservation,* 4, 1(1959).
8. K. Yamasaki. *Shôsôin Medicinals* * ed. Y. Asahina, (Osaka, 1955), p. 481.
9. K. Yamasaki, *Bukkyô Geijutsu* ("Ars Buddhica"),* No. 31, 62(1957).
10. K. Yamasaki: *Wall Painting of Daigoji Pagoda* * ed. O. Takata (1959).
11. T. Akiyama, *Bijutsu Kenkyû,** No. 168, 17(1952).
12. K. Yamasaki and H. Nakayama, *Bijutsu Kenkyû,** No. 174, 47(1954).
13. T. Asahina, F. Yamasaki and K. Yamasaki. "UNESCO International Conference on Radio-isotopes in Scientific Research," (Paris, 1957), Research Paper No. 64.
14. R. Uemura, T. Kameda, K. Kitamura and K. Yamasaki, "Bulletin of Archives and Mausolea Division," * Imperial Household Agency, No. 4, 68(1954).
15. K. Yamasaki, *Bijutsu Kenkyû,** No. 212, 31(1960).
16. R. B. Kahn et al., *Oil & Color Chemists Assoc.,* 32, 24(1949); Freer Abstracts, No. 1038.
17. R. J. Gettens, *Technical Studies,* 6, 186(1938).
18. R. J. Gettens, *Technical Studies,* 6, 281(1938).
19. K. Yamasaki: unpublished results.
20. A. Stein: *Serindia,* Vol. III, Appendix D; Unpublished results of K. Yamasaki.
21. L. Warner: *Buddhist Wall Paintings,* 1938, Chapter III, p. 10. (Analytical data by Gettens.)
22. K. Yamasaki: *Bijutsu Kenkyû,** No. 153, 43(1949); J. Tamura and Y. Kobayashi: *Ching-Ling,* Kyoto University, 2 vols. 1953.
23. P. I. Kostrov and E. G. Sheinina, *Studies in Conservation,* 6, 90(1961).

* In Japanese.